# Metallographic Etching

## Second Edition

### Techniques for Metallography Ceramography and Plastography

**Günter Petzow**

ASM INTERNATIONAL

ASM International®
Materials Park, Ohio 44073-0002
www.asminternational.org

First printing, November 1999
Second printing, December 2000
Third printing, July 2008
Digital printing, April 2014
Fifth printing, April 2019
Sixth printing, December 2019
Seventh printing, November 2021
Eighth printing, November 2025



Library of Congress Cataloging-in-Publication Data
[Metallographisches Ätzen, English]
Metallographic etching/by Günter Petzow, in collaboration with Veronika Carle,
Translated by Uta Hanisch.—2nd ed.
Originally published: Metals Park, Ohio: American Society for Metals, c1978.
Includes bibliographical references and index.
1. Metallographic specimens. 2. Metals—Etching. I. Title.
TN690.P4713 1999 669'.95'0282—dc21 99-29323

ISBN-13: 978-0-87170-633-1 (print)
ISBN-10: 0-87170-633-4 (print)
ISBN: 978-1-61503-220-4 (pdf)
SAN: 204-7586

ASM International®
Materials Park, OH 44073-0002
www.asminternational.org

Printed in the United States of America

Cover Image: "Deeply Etched Graphite Yellowstone," submitted by Amir Ardalan Rezaie and Reza Abbashian, University of California, Riverside. 2018 International Metallographic Contest, First Place, Class 4, Artistic Microscopy, Color

# Dedication

I wish to dedicate this edition of the Etch book to the International Metallographic Society (IMS) and especially to all its fellow members who have contributed with their friendly cooperation over the decades to the development of modern metallography.

# Contents

**Preface to the German Edition** . . . . . . . . . . . . . . . . . . . . . . . . . . . . . . . . . . . **VIII**
**Preface** . . . . . . . . . . . . . . . . . . . . . . . . . . . . . . . . . . . . . . . . . . . . . . . . . . . . . . **IX**
**Introduction** . . . . . . . . . . . . . . . . . . . . . . . . . . . . . . . . . . . . . . . . . . . . . . . . . . **1**
**1 Metallography** . . . . . . . . . . . . . . . . . . . . . . . . . . . . . . . . . . . . . . . . . . . . . . **7**

1.1 Preparation of Metallographic Specimens . . . . . . . . . . . . . . . . . . . . . . 7
1.2 Specimen Sectioning . . . . . . . . . . . . . . . . . . . . . . . . . . . . . . . . . . . . . 8
1.3 Mounting . . . . . . . . . . . . . . . . . . . . . . . . . . . . . . . . . . . . . . . . . . . . . 11
1.3.1 Clamping . . . . . . . . . . . . . . . . . . . . . . . . . . . . . . . . . . . . . . . . . . . 12
1.3.2 Embedding . . . . . . . . . . . . . . . . . . . . . . . . . . . . . . . . . . . . . . . . . 13
1.3.3 Special Mounting Techniques . . . . . . . . . . . . . . . . . . . . . . . . . . . . 15
1.4 Identification (Marking) . . . . . . . . . . . . . . . . . . . . . . . . . . . . . . . . . . 18
1.5 Grinding and Polishing . . . . . . . . . . . . . . . . . . . . . . . . . . . . . . . . . . . 18
1.5.1 Mechanical Grinding and Polishing . . . . . . . . . . . . . . . . . . . . . . . . 19
1.5.2 Microtome Cutting and Ultramilling . . . . . . . . . . . . . . . . . . . . . . . 29
1.5.3 Electrolytic Grinding and Polishing . . . . . . . . . . . . . . . . . . . . . . . . 30
1.5.4 Chemical Polishing . . . . . . . . . . . . . . . . . . . . . . . . . . . . . . . . . . . 34
1.5.5 Combined Polishing Methods . . . . . . . . . . . . . . . . . . . . . . . . . . . . 34
1.5.6 Evaluation of Polishing Methods . . . . . . . . . . . . . . . . . . . . . . . . . . 36
1.6 Cleaning . . . . . . . . . . . . . . . . . . . . . . . . . . . . . . . . . . . . . . . . . . . . . 37
1.7 Metallographic Etching . . . . . . . . . . . . . . . . . . . . . . . . . . . . . . . . . . . 38
1.7.1 Optical Etching . . . . . . . . . . . . . . . . . . . . . . . . . . . . . . . . . . . . . . 39
1.7.2 Electrochemical (Chemical) Etching . . . . . . . . . . . . . . . . . . . . . . . 40
1.7.3 Physical Etching . . . . . . . . . . . . . . . . . . . . . . . . . . . . . . . . . . . . . 44
1.7.4 Etching Reproducibility . . . . . . . . . . . . . . . . . . . . . . . . . . . . . . . . 45
1.7.5 Etching Terminology . . . . . . . . . . . . . . . . . . . . . . . . . . . . . . . . . . 47
1.7.6 Definitions of Etching Terms . . . . . . . . . . . . . . . . . . . . . . . . . . . . 47
1.8 Field Metallography/Nondestructive Metallography . . . . . . . . . . . . . . . 50
1.9 Preparation and Etching Recipes for Metallic Materials . . . . . . . . . . . . 51
1.9.1 Silver . . . . . . . . . . . . . . . . . . . . . . . . . . . . . . . . . . . . . . . . . . . . . 53
1.9.2 Aluminum . . . . . . . . . . . . . . . . . . . . . . . . . . . . . . . . . . . . . . . . . . 57
1.9.3 Gold . . . . . . . . . . . . . . . . . . . . . . . . . . . . . . . . . . . . . . . . . . . . . . 64

1.9.4 Beryllium . . . 67
1.9.5 Bismuth and Antimony . . . 70
1.9.6 Cadmium, Indium, and Thallium . . . 72
1.9.7 Cobalt . . . 75
1.9.8 Refractory Metals (Chromium, Molybdenum, Niobium, Rhenium, Tantalum, Vanadium, and Tungsten) . . . 80
1.9.9 Copper . . . 88
1.9.10 Iron, Steel, and Cast Iron . . . 94
1.9.11 Semiconductors (Germanium, Silicon, Selenium, Tellurium, $A_{III}B_{VI}$, and $A_{II}B_{VI}$ Compounds) . . . 102
1.9.12 Hafnium and Zirconium . . . 109
1.9.13 Mercury (Amalgams) . . . 112
1.9.14 Magnesium . . . 114
1.9.15 Manganese . . . 117
1.9.16 Nickel . . . 118
1.9.17 Lead . . . 125
1.9.18 Platinum Metals (Platinum, Palladium, Rhodium, Iridium, Ruthenium, and Osmium) . . . 130
1.9.19 Plutonium, Thorium, and Uranium, Americium, and Neptunium . . . 133
1.9.20 Tin . . . 135
1.9.21 Titanium . . . 139
1.9.22 Yttrium, Lanthanum, and Rare Earth Metals . . . 144
1.9.23 Zinc . . . 148

**2 Ceramography . . . 153**

2.1 Sample Preparation for Ceramic Materials . . . 153
2.2 Sampling . . . 154
2.3 Mounting . . . 154
2.4 Identification . . . 155
2.5 Grinding and Polishing . . . 155
2.5.1 Mechanical Grinding and Polishing . . . 155
2.5.2 Other Grinding and Polishing Methods . . . 158
2.6 Cleaning . . . 158
2.7 Ceramographic Etching . . . 159
2.7.1 Optical Etching . . . 159
2.7.2 Electrochemical (Chemical) Etching . . . 160
2.7.3 Physical Etching . . . 162
2.8 Preparation and Etch Recipes of Ceramic Materials . . . 163
2.8.1 Oxide Ceramics . . . 164
2.8.2 Nitride Ceramics . . . 168
2.8.3 Boride Ceramics . . . 169

2.8.4 Carbide Ceramics . . . 169
2.8.5 Phosphide and Sulfide Ceramics . . . 172
2.8.6 Hardmetals (Cemented Carbides) . . . 172

**3 Plastography . . . 175**

3.1 Plastographic Sample Preparation . . . 177
3.2 Plastographic Etching . . . 179
3.3 Etchants for Polymers . . . 180
3.4 Plastographic Thin Sectioning . . . 181
3.5 Plastographic Microtome Sectioning . . . 182
3.6 Light Microscopic Studies of Thin Sections or Microtomed Sections . . . 183

**4 Appendix . . . 185**

4.1 Designation of Etchants . . . 185
4.2 List of Chemicals . . . 190
4.3 Hazardous Chemical Symbols and Their Interpretation . . . 196
4.4 Handling Tips for Hazardous Materials . . . 198
4.4.1 Rules, Tips, and Precautions . . . 198
4.4.2 Important Special Precautions . . . 199
4.4.3 Dangerous Reactions of Chemicals . . . 202
4.4.4 Disposal of Hazardous Materials . . . 203

**5 Literature References . . . 205**

Journals . . . 205
General Textbooks and Reviews . . . 205
Handbooks, Compilations, and Tables . . . 208
Metallographic Preparation . . . 209
Metallic Materials . . . 210
Ceramic Materials . . . 211
Polymer Materials . . . 212

# Preface to the German Edition

It is a true coincidence that the time lapse between the edition of this book and the previous one is approximately the same (18 years) as the one between the last edition and the first one. During that time, the content of *Metallographic Etching* had become noticeably more extensive and had to be reprinted several times with translations into English, Japanese, Chinese, Korean, and Rumanian.

One could nearly quote the preface from the Fifth Edition for this Sixth Edition, since once again considerable progress in the materials field and its testing methods has taken place. New developments of all types of materials and novel material combinations for high-tech applications have created a scenario to which the metallographic analysis methods had to be adapted as well. The multitude of materials and their applications led to an overwhelming wealth of new procedures that were often replacing the older ones. This fact has been considered in the present edition of this book. For example, metallographic preparation, specific for a certain material before etching, has been extended. The chapter dealing with ceramography has also been adapted to the most recent developments within the past years, and a chapter on plastography has been added. Descriptions of very specific, risky, and complicated recipes have been omitted to include easier and safer ones that can be followed even in less well equipped laboratories.

I am very grateful to all who enthusiastically contributed discussions and consultations to this revised version of the book. Very special thanks go to my colleagues from the Metallography Department of the Powder Metallurgical Laboratory—V. Carle, F. Predel, U. Täffner, and U. Schäfer—who have assisted with great diligence in the literature reviews and with the validation of the recipes; thanks are also due to G. Weilnhammer, R. Habermann, G. Kiessler, and H. Wendel. The numerous critical comments and valuable observations made by R. Telle, G. Schneider, G. Petermann, M. Schwarz, C. Herget, E. Sachowa, and E. Tenckhoff are greatly appreciated. I am also very grateful to Messrs. G. Kaiser and J. Dahmen, who contributed immensely to the extensive task of the chemistry index and the safety aspects, as well to my diligent secretary, D. Wieder, who managed the outlay of the manuscript most efficiently.

Stuttgart
January 1994
Günter Petzow

# Preface

One of the most important prerequisites for reliable results through microscopic analysis is the careful and reproducible preparation of the specimens to be analyzed. They are often metallographically prepared specimens for reflective light microscopy, a challenge that is well covered with the content of this book.

Compared to the first edition, in 1978, the present edition has been considerably expanded. The new developments in the metallic materials groups as well as novel material combinations as they are required for many fields of high technology have been covered in the "Metallography" chapter. The chapter "Ceramography" deals equally with the rapid progress during the past few years. The chapter on "Plastography" is a new addition. Very specific, high-risk, and complicated recipes have not been included in these chapters. Instead, simpler and more basic recipes that are easily reproducible in laboratories with less sophisticated equipment are described.

The book has already been found to be a valuable basis for many materialographical workshops in Germany, England, Brazil, Japan, and the United States. Of course, I have to admit that a complete and finite version covering all of the multifaceted and rapidly changing sectors of materials science is impossible. I am therefore grateful for and will appreciate any suggestion by the reader for supplement and improvement of the edition.

I wish to thank my colleague Uta Harnisch of Alcan Ltd. for the preparation of this second edition of *Metallographic Etching*. She translated this edition with stamina and concern and has therefore proven that traditional teamwork exists within IMS. Her commitment and technical knowledge helped to realize and firmly advance the translation of this edition.

I wish to thank Jim Nelson and George Vander Voort of Buehler Ltd. for the critical review of this manuscript and Veronica Flint of ASM for her attentive preparation of this book.

Stuttgart
March 1999
Günter Petzow

# Introduction

Modern technology relies heavily on the use of engineering materials. Their structure and properties are determined by many features, extending from atomic distances measured in nanometers, to the final product, measured in centimeters or even meters (Fig. 1). This size scale, spanning several orders of magnitude, begins with the crystal structure (lattice structure) and includes configurations in the nano, micro, and macro range, all of which contribute to the broad spectrum of properties.

It follows that the internal structure of materials is set by the atomic species and their specific three-dimensional arrangement. The structural architecture is determined by symmetry, but in reality deviations from perfect symmetry (*lattice*

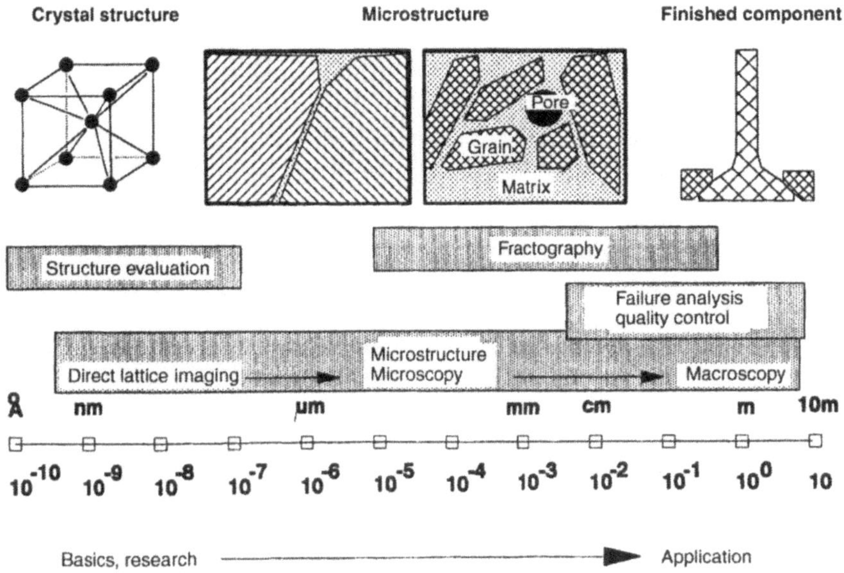


**Fig. 1** Spectrum of dimensions from crystal structure to finished component

*defects*) are quite common; their type, size, shape, quantity, and distribution, together with the perfect structural regions, establish the so-called *microstructure*. Sometimes the microstructure is defined by the perfect structural regions only, according to their type, size, shape, quantity, and distribution. However, the definition that includes lattice defects is more realistic by far, since it can differentiate between defects that may have a variety of effects on the properties. Table 1 describes the lattice defects according to their geometrical extension.

Usually these descriptions of the microstructure are sufficient for practical applications. In more detailed descriptions of the microstructure, one differentiates between those lattice defects that are in a state of thermodynamic equilibrium and those that are not. *Only the nonequilibrium lattice defects are considered part of the microstructure.* They comprise all those lattice defects that increase the energy content of the crystals and can therefore be eliminated by appropriate heat treatments, sometimes requiring extremely long times. This applies to all lattice defects. However, vacancies, antiphase boundaries, and stacking faults may also be present in thermodynamic equilibrium. They are known therefore as *structural lattice defects* that serve to stabilize the crystal structure and for that reason are they categorized to the equilibrium lattice structure. A hypothetical, *ideal single crystal* in thermodynamic equilibrium does not have a microstructure as such, but only a crystal structure. However, a *real single crystal* does have a microstructure, even though it lacks the most prominent microstructural characteristics, the grain boundaries.

A definition of the microstructure that accounts for these more academic views is:

> The microstructure is characterized by the type, size, shape, quantity, and distribution of lattice defects that are not in a thermodynamic equilibrium.

The material properties are related to both the constituent phases and the microstructure, that is, the totality of lattice defects in the nonequilibrium state. This effect of the microstructure is particularly significant with regard to certain properties, for example, strength, thermal conductivity, corrosion resistance, and diffusion rate. These properties can be described as *microstructure-sensitive,* in

**Table 1 Lattice defects listed according to geometric extension**

| Zero dimensional | One dimensional | Two dimensional | Three dimensional |
|---|---|---|---|
| Vacancies | Dislocations | Stacking faults | Foreign particles, precipitates |
| Interstitial atoms | | Antiphase domain boundaries | Pores, shrinkage cavity |
| Complex point defects | | Low-angle and high-angle grain boundaries, phase boundaries | |

contrast to *microstructure-insensitive,* properties such as Young's modulus, density, melting point, transformation temperature, and thermal expansion coefficient. Thus, the microstructure affects some very important properties and is therefore given considerable attention in the science, development, and testing of materials. The greater the demands on a material, the more sophisticated the microstructural input becomes, that is, the more exacting the requirements on the microstructural parameters will be. The objective is an optimum adjustment of the microstructure to fit a specific property profile; "microstructural engineering" or "microstructural design" are the key words that characterize these problems. Only in this way can tight material tolerances be maintained for components that are used in highly critical applications. The examination of materials and their microstructures therefore takes an important place in materials research and applications and includes all materials, for example, metals, ceramics, and polymers, as well as their alloys and composites.

Among the methods of microstructural examination, *materialography* (metallography) plays an outstanding role.

> Materialography (metallography) is defined as an investigative method of materials science. It encompasses the optical examination of microstructures, and its goal is a qualitative and quantitative description of the microstructure.

The term "metallography" does not literally apply to nonmetallic materials. The term "materialography" has already been suggested and would be more appropriate in this context. However, to date this terminology has not been fully accepted, possibly because the significance of the relationship between microstructures and properties was first recognized in metallic materials. Accordingly, metallographic techniques and methods were first developed and were then adapted with certain modifications to nonmetallic materials. The term "metallographic examination" is frequently and imprecisely applied to the examination of ceramics and polymers, instead of using the more accurate terms of "ceramographic" and "plastographic" examination.

The optical methods used in metallographic practice are basically the techniques of light optical and electron optical microscopy. Visual light, with its wavelengths ranging from 400 to 800 nm, has a resolution of approximately 300 nm.* The resolution of an electron beam, on the other hand, is considerably higher.** It is directly related to the accelerating voltage and is 0.3 nm at 100 kV.

---

*nm = $10^{-3}$ μm = $10^{-6}$ mm = $10^{-9}$ m = 10 Å

**Two adjacent points can still be resolved when their distance is $d = \lambda / n \times \sin \alpha$, where $\lambda$ = wavelength of the light, $n$ = refractive index of the medium, and $\alpha$ = angle of incidence.

Figure 2 shows the magnification and resolution obtained with common optical equipment. It also indicates the extent of microstructural constituents.

A comparison of the magnifications or resolutions obtained from the various equipment with which the microstructures can be observed shows that the entire range of matter within a microstructure can be made visible, that is, can be evaluated metallographically.

> A complete and continuous description of a microstructure can be achieved by employing macroscopic and light optical and electron optical techniques.

Due to the many different physical and technical requirements for the various optical methods, the handling of the equipment, the specimen preparation, and the interpretation of micrographs is complex. A full range of equipment, from a magnifying glass to light microscopes and scanning electron microscopes, as well as the high-resolution and high-voltage electron microscopes, are available. Thought must be given to the selection of the appropriate investigative technique that generates the desired microstructural information at a minimum effort.

A very important requirement for all optical examinations of the microstructure is reliable specimen preparation. This includes the "as-received" specimens, sections, thin sections, and foils (Fig. 2). Numerous flaws and

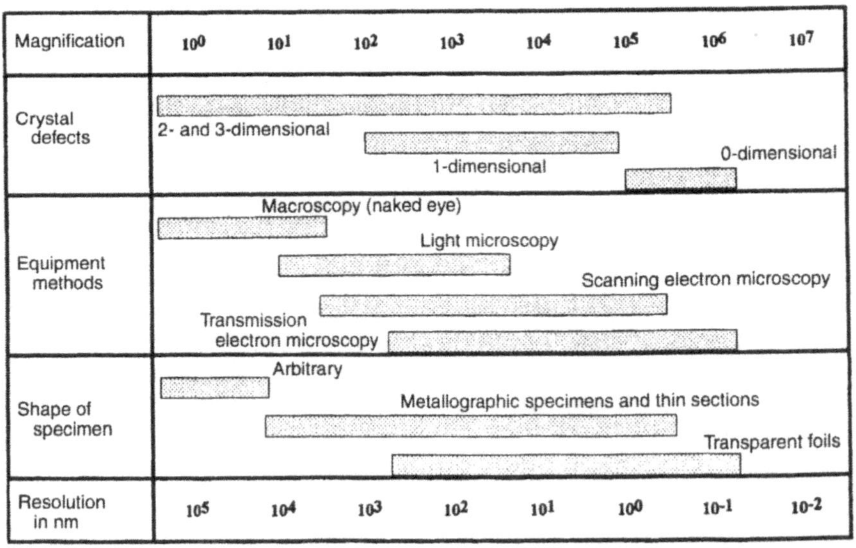


**Fig. 2** Crystal defects, equipment and specimen shape used for the examination of microstructural features with corresponding ranges of magnification and resolution

imperfections can develop during metallographic preparation. Improperly prepared samples can lead to misinterpretation of the microstructure and consequently lead to incorrect material developments to inappropriate measures for process control, or to misinterpretations in failure analysis.

> Without proper specimen preparation, the result of any microstructural investigation will remain questionable.

In routine metallographic work, the investigation of sections is a dominant feature. The metallographic section preparation is very intricate and to a large degree empirical. Great care, skill, and attention to detail are required. The following chapters contain important suggestions for the specimen preparation of metals, ceramics, and polymers.

Although different structural features in different materials species are addressed, many similarities exist in their specimen preparation, in particular with respect to metallographic sections.

# 1 Metallography

The prominent position that technology holds can hardly be imagined without the presence of metallic materials. Metals and their alloys have nearly without exception had a direct or indirect impact on all technological developments. Like no other materials group, they cover an extensive spectrum of applications with their enormous variety of alloys and their related range of properties. The number of metallic materials extends to several thousand and is continuously growing to meet the latest requirements of technology. The luster and the good electricity and heat conductivity, as well as the high ductility and strength, combine to provide characteristic and unique properties.

Many properties of metallic materials, such as yield strength, elongation, ultimate tensile strength, coercivity, thermal conductivity, and corrosion resistance, as well as the electric resistance and coefficient of diffusion, are more or less directly related to the microstructure; they have a high microstructural sensitivity. The understanding of the relationship between microstructure and properties therefore plays an important role in the control and development of metallic materials. The examination of the microstructure, *metallography,* is thus an important test method during production and a very powerful tool for detecting fabricating defects and the causes of material failures. Without doubt, most investigations are carried out with incident light microscopy to reveal the various microstructural features. Also, it is obvious that an important prerequisite is a well-prepared specimen.

## 1.1 Preparation of Metallographic Specimens

The methods of preparing metallographic sections for macroscopic and microscopic investigations are numerous and diverse. This is due to the variety of materials requiring investigation and the manner in which we have inherited much of our current data. Handed-down formulas must be considered, as well as patent rights, commercial aspects, and modern developments or improved preparation techniques and equipment. Thus, a comprehensive survey of specimen preparation is at best difficult. Nevertheless, some correlations exist and are systematically explained prior to the tabulation of currently accepted metallographic etchants. Unfortunately, there is no universal technique that meets all the demands of metallographic specimen preparation.

Metallographic preparation usually requires a specific sequence of operations that includes sectioning, mounting, identification, grinding, polishing, cleaning, and etching. Each of these steps can be carried out in different ways and may vary according to the specific material properties. In principal, specimen preparation requires several steps, even though not all need to be pursued in every application. Great care must be taken in performing each individual step because carelessness at any stage may affect the later steps. In extreme cases, improper preparation may result in a distorted structure leading to erroneous interpretation.

> A satisfactory metallographic specimen for macroscopic and/or microscopic investigation must include a representative plane surface area of the material. To clearly distinguish the structural details, this area must be free from changes caused by surface deformation, flowed material (smears), plucking (pullouts), and scratches. In certain cases, the edges of the specimen must be preserved.

By observing simple commonsense principles, an acceptable preparation is possible for any solid-state material, although in many cases it would require much patience. Even for routine examinations and for the least-critical applications, poor specimen preparation is unacceptable because the observations and the resulting conclusions are, at best, questionable.

## 1.2 Specimen Sectioning

The first step in specimen preparation—the selection and separation of samples from the bulk material (sampling)—is of special importance. If the choice of a sample is not representative of the material, it cannot be corrected later. It is also difficult to compensate later for improper sectioning, because additional, time-consuming corrective steps are necessary to remove the initial damage.

> Sectioning should render a plane surface for the following preparation without causing critical changes in the material.

Alterations in the microstructure of the specimen can be produced by deformation and the creation and further development of cracks and breakouts. Due to heat generation, recrystallization, local tempering, and in extreme cases, partial melting may occur. These problems can be minimized by the use of generous amounts of inert lubricants and coolants (water, oil, compressed air, etc.).

The sectioning techniques are summarized in Fig. 1.1 and arranged according to the different sectioning mechanisms. When sectioning with a torch or by normal mechanical sawing, cutting, sand blasting, or cleaving, care must be taken to cut sufficiently far from the area of interest to avoid harmful effects. The prespecimen (initial specimen) must be large enough for the final sample to remain in its original state. This prespecimen may then be heavily ground or cut by more sophisticated and delicate means to produce the desired plane for further preparation. Ideally, only methods that produce surfaces suitable for immediate fine grinding or even polishing should be used. Such methods include abrasive cutting, ultrasonic chiseling, arc cutting, and electrochemical machining. Although the goal is to use such material-preserving methods for sectioning in the first place without the detour of a prespecimen, these methods are troublesome and time consuming and only useful for special applications (single crystals, semiconductors, and brittle materials). They are not economical for routine work.

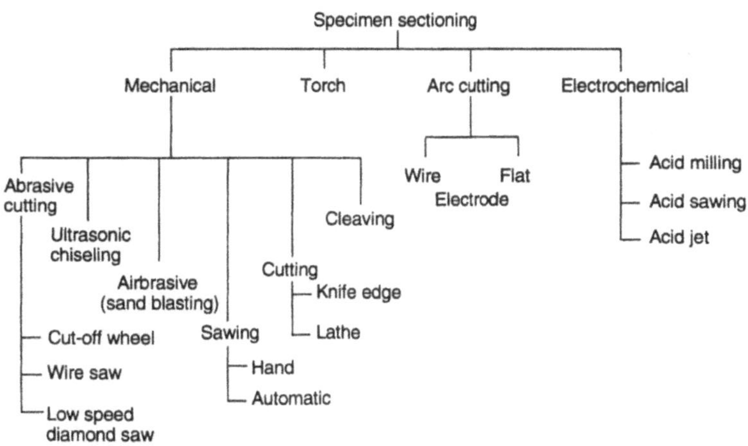


**Fig. 1.1** Methods of metallographic specimen sectioning

The most versatile and economical sectioning method is abrasive cutting. A thin, rapidly rotating, consumable abrasive wheel produces high-quality, low-distortion cuts in times ranging from seconds to several minutes, depending on the material and the cross-sectional area. This technique is almost universally applicable. Important parameters in abrasive cutting are wheel composition, coolant condition, and technique.

**Abrasive wheels** consist of abrasive grains (alumina, silicon carbide, boron nitride, diamond) bonded together with either resin or rubber or a rubber-resin combination, or metal.

The abrasive grains become rapidly worn out during cutting of hard materials and must be continuously replaced by newly exposed grains. When cutting hard materials, a cutting wheel with a soft binder (soft cutting wheel) is chosen to promote a fast removal of used grains and the continuous exposure of new abrasive grains, thus always maintaining a sharp cutting edge. On the other hand, a cutting wheel with a hard binder (hard cutting wheel), is recommended for soft materials, since a fast exchange of abrasive particles is not necessary. Although soft cutting wheels are more rapidly consumed than hard cutting wheels, they are more suitable for harder materials, since their cutting action is considerably faster because of the renewed exposure of fresh abrasive grains.

Other criteria for the selection of suitable abrasive wheels are the grain size and the concentration of the abrasive particles, as well as the thickness of the wheel. The concentration depends on the special cutting task and on the material to be sectioned, the binder, and the abrasive grain size. The abrasive concentration determines the removal rate and the durability of the abrasive wheel. A wheel with a high concentration and coarse abrasive particles is generally recommended for small areas of contact, while a cutting wheel with low concentration is typically used for large areas of contact. Depending on the type of material to be sectioned, cutting wheels of different compositions should be used, and their selection is dictated by hardness and ductility of the material to be cut. Table 1.1

**Table 1.1 Application of cutting wheels**

| Materials | Cutting wheel: abrasive/binder |
|---|---|
| Steel, ferrous materials, hardened steels | $Al_2O_3$ (corundum)/bakelite |
| High-alloy steels | Cubic boron nitride (CBN)/bakelite |
| Nonferrous metals, hardmetals | Silicon carbide (SiC)/bakelite |
| Hard and tough materials, cermets, ceramics | Diamond/bakelite |
| Hard and brittle materials, ceramics, minerals | Diamond/metal |

lists some common materials and their appropriate cutting-wheel/binder combinations.

**Cutting wheel machines** are available with high or low speed, with or without a feeding device, and even with precisely guided sample holders. The machines must always be equipped with a cooling system to prevent excessive heat that might affect the microstructure of the specimen; the coolant is ordinarily water, to which a corrosion inhibiting agent can be added.

## 1.3 Mounting

Mounting specimens in a holding device is necessary when preparing irregular, small, very soft, porous, or fragile specimens, and in those cases where edge retention is required. Embedding is indispensable when multiple specimens are to be included in a single mount or when automatic equipment is to be used in the following preparation. In most cases, mounting follows sectioning, but in the handling of a great number of very small specimens, it may be advantageous to reverse this order. In general, the mounting procedure can be easily adapted to the special problem in question. The shape, size, and numbers as well as the hardness, brittleness, porosity, and heat and pressure sensitivity of the specimens have to be considered when mounting. Other considerations are: should a cross or a longitudinal section be prepared, is a controlled material removal required, is good edge retention needed, and should the preparation be carried out manually or with automatic equipment in specific sample holders.

A suitable *mounting media* must meet several criteria: it must have good adhesion to the specimen, sufficient mechanical strength (hardness), and chemical resistance to etchants or solvents that are used during the preparation. For electrolytic polishing, scanning electron microscopy examination, or microprobe analysis, the mounting medium has to be electrically conductive. The mounting material should be easy to handle, economical, if necessary easy to remove, and it should not affect the specimens. For some investigations, a transparent mounting medium is more appropriate than an opaque material, and in cases where the specimens have to be analyzed with x-rays, a mounting material free of any interference reflexes should be selected.

Because of these varied requirements, many different techniques were developed for the mounting and embedding of metallographic samples. They are summarized schematically in Fig. 1.2 and can be described as two basic types of mounting: *clamping* with a sample holder or clamp, and *embedding* the specimens in organic or inorganic materials.

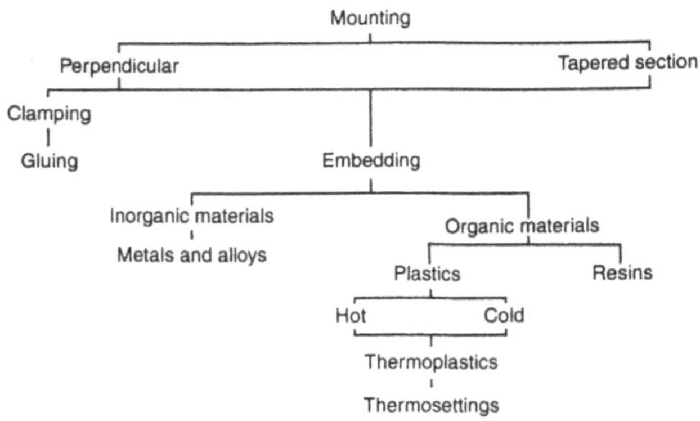


**Fig. 1.2** Methods of metallographic mounting

## *1.3.1 Clamping*

Clamping is a traditional, simple, and inexpensive technique; several examples are shown in Fig. 1.3. The clamps are usually made of soft steels, stainless steels, aluminum alloys, or copper alloys. Stainless steel is the best choice for a wide variety of materials. The sample and clamp material have to be compatible to maintain similar material removal during the grinding and polishing steps and to prevent occurrence of "ghost" microstructures during etching.

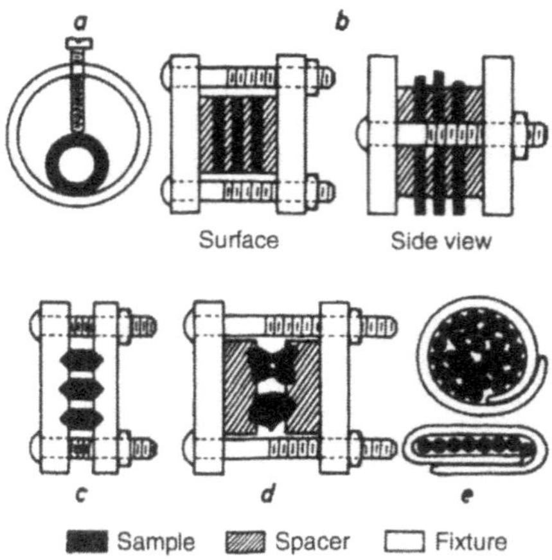


**Fig. 1.3** Examples of different fixtures used for clamping. (a) Tube section. (b) Sheet specimens. (c) Rods. (d) Irregular pieces. (e) Wires

Elastic spacers (e.g., cork, PVC, styrofoam, or rubber) are useful for reducing the compression from the clamping onto the specimens and reduces deformation. These spacers may interfere with the sample preparation, causing gaps to retain solvents and grinding and polishing media that may be difficult to remove and can lead to artifacts.

## *1.3.2 Embedding*

Embedding or casting of plastic materials around the specimens is the most popular technique and can be divided into "cold" and "hot" mounting, depending on whether or not heat is needed for the polymerization process (Fig. 1.4).

**Cold mounting** (room-temperature curing) requires the mixing of two agents (a crude polymer and a catalyst); this mixture is then cast over the specimen within a mold, in which it reacts to form a solid part.

A slightly higher pressure during curing improves the adhesion to the specimen. Special equipment is available to mount several samples simultaneously.

**Hot mounting** (compression molding) requires a mounting press where the sample and the mounting compound can be heated and simultaneously compressed.

Two essential types of mounting materials are available, *thermosettings* and *thermoplastics*. Both types are available as hot and cold mounting compounds, depending on whether the polymer reaction occurs with added heat or with the addition of a catalyst.

The curing of thermosetting materials is irreversible, and they cannot be resoftened after curing; cured thermoplastic materials, however, can be remelted again at elevated temperatures. Tables 1.2 and 1.3 list information on hot and cold mounting media and their properties.

Important requirements of hot and cold mounting materials are:

- Specimens thoroughly cleaned either with an ultrasonic cleaner or lightly brushed with water containing a detergent

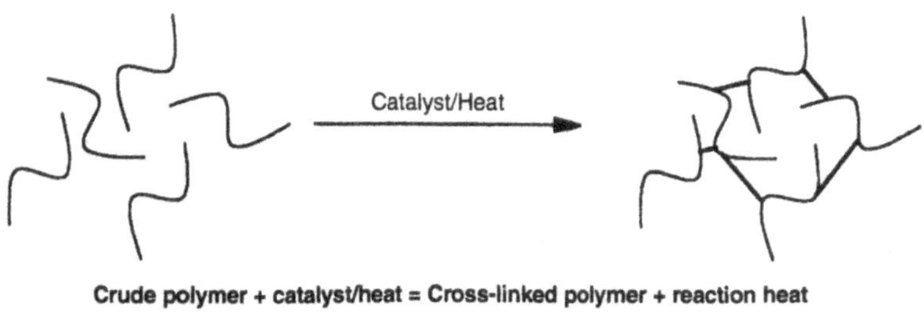


**Fig. 1.4** Schematic of the polymerizing process during cold and hot mounting

## Table 1.2 Comparison of hot and cold mounting materials

| Hot compression mounting materials | Cold/room-temperature-curing mounting materials |
|---|---|
| Powder, granulates, or preforms are densified by pressure and heat in a mounting press. | Liquid and/or powder are mixed together with a hardener and cast into suitable mold. |
| Starting material has a durable shelf life. | Starting material should be stored cool and has a limited shelf life. |
| Time required per mount is ~15 min. | Curing time per sample ranges from 10 min to 12 h; many samples can be mounted simultaneously. |
| **Thermosetting plastics** | |
| Phenol resins (Bakelite)<br>Epoxy resins<br>Diallyl phthalate | Epoxy resins<br>Polyester resins |
| Polymerization is irreversible; material cannot resoften.<br>Heat to ~150 °C under pressure.<br>Samples can be removed hot from the mounting press, but cooling under pressure is recommended. | Polymerization is irreversible; material will not resoften.<br>Hardening reaction may increase the temperatures! This is directly related to the mixing ratio, the ambient temperatures, the volume of the mixed components, and the heat transfer by the molds. |
| **Thermoplastics** | |
| Acrylics<br>Material can be resoftened with heat.<br>Heat without pressure and cool under pressure. | Acrylics<br>Material can be resoftened. Temperature may increase from 50–120 °C. Curing times are short. |

## Table 1.3 Properties of some important mounting materials

| Material | Property |
|---|---|
| **Hot mounting materials** | |
| Phenol resins (bakelite) | Low hardness, poor adhesion (may be improved during cooling under pressure). Poor chemical resistance to aggressive chemicals and hot etchants. Easy to use, low cost |
| Epoxy resins | Only little shrinkage during curing, good edge retention. Resistant to etchants |
| Diallyl phthalate | Suitable for hard materials. No shrinkage during curing. Resistant to aggressive chemicals and hot etchants. Mounting conditions must be strictly followed. |
| Acrylics | Care must be taken during grinding; material may crack due to imposed stresses. Poor adhesion. Not resistant to aggressive chemicals. Transparent. Sample should be well cooled during curing. Suitable for pressure-sensitive specimens; pressure is only to be applied during the cooling cycle. |
| **Cold mounting materials** | |
| Epoxy resins | Good adhesion. High viscosity, fills cracks, gaps, and pores easily and is therefore well suited for infiltration. Resistant to etchants and solvents. Nearly transparent. Mold material should be made of silicon rubber, polyethylene, or Bakelite. Curing time at least 8 h. Work under a fumehood because poisonous fumes are being generated. Skin irritant |
| Polyester resins | Good abrasion resistance, therefore well suited for hard materials. Shrinkage. Chemical resistance varies with the product. |
| Acrylics | Shrinkage. Short curing times. Poor resistance to alcohol and chlorohydrocarbon |

- Inert to the specimen, mold, and etchant
- Moderate viscosity during the actual embedding process; no generation of air bubbles after solidification
- Low linear shrinkage and good adhesion to the specimen
- Grinding and polishing behavior similar to those of the specimen, as well as similar chemical resistance to all reagents used during the specimen preparation

These requirements are satisfied by many commercially available plastic materials for the two groups of thermoplastic and thermosetting materials. When selecting a suitable mounting material, it is helpful to contact the supplier for information about specific applications and the properties of a certain mounting material.

Heating as it occurs during hot mounting can also take place during the cold mounting process. Due to exothermic reactions, the temperature may rise to above 150 °C (300 °F). However, this temperature increase can be controlled by lowering the catalyst addition or by cooling the mount appropriately with chilled water or a flow of compressed air.

> Materials that are temperature and pressure sensitive have to be cold mounted. When mounting only a few samples, hot mounting is more time efficient than cold mounting; the reverse is true for a larger number of samples.

To support wires during mounting, various metal or plastic clips are commercially available. They do not react with the different mounting media and are resistant to most common etchants.

### *1.3.3 Special Mounting Techniques*

**Taper Technique.** Thin layers, platings, or diffusion zones present a problem due to the difficulty in observing and measuring the thin areas. The apparent thickness of thin layers may be increased by means of a taper technique, in which the specimen is tilted at a shallow angle in the mold as shown in Fig. 1.5(a) and (b). The width advantage gained is dependent on the angle, as tabulated in Table 1.4.

When the thickness of the coated sample is known (Fig. 1.5a), then the following equation applies: $L = S \cdot L'/S'$, or if the angle $\alpha$ (Fig. 1.5b) of the supporting wedge is known, then $L = L' \cdot \sin \alpha$.

**Sample Preparation for Edge Retention.** When analyzing thin layers and surface defects, it is important to have a sample with good edge retention. The contact between the sample and the mounting material has to be optimum. Rinsing the sample with a wetting agent and using a material generating low shrinkage during the curing cycle is beneficial. Another technique can be used in which the sample surface is reinforced with an additional layer; for example, the sample can be galvanized or an electroless coating can be deposited. Equipment and electrolytic solutions are available for the deposition of layers from thicknesses of 10 μm and more. Nonconductive materials can also be plated.

By addition of ceramic particles, such as alumina powder, the hardness of the mounting material can be increased and adjusted to the hardness of the sample, therefore resulting in a similar grinding and polishing behavior of the mounted unit and a better edge preservation.

**Mounting for Electrolytic Sample Preparation.** For electrolytic polishing or etching, the metallographic sample must have an electric contact. This can either be a hole drilled through the mounting material or a wire attached to the sample before mounting. Conductive mounting media are also commercially available.

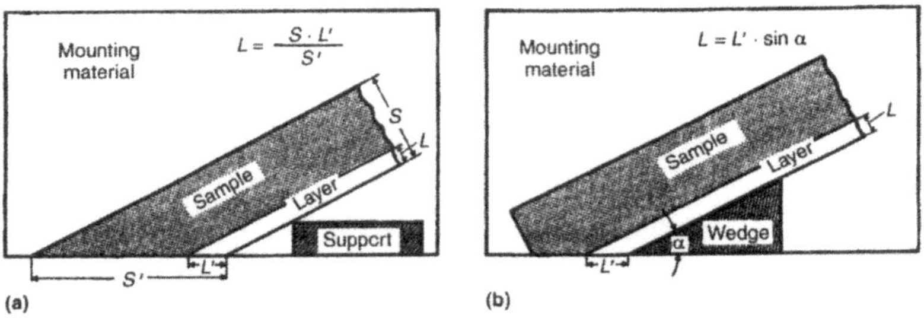


**Fig. 1.5** Taper sectioning (oblique mounting to increase the width of a layer). *S*, sample thickness; *S'*, sample thickness of the taper section; *L*, layer thickness; *L'*, layer thickness of the taper section; α, tilt angle

**Table 1.4 Increase in width of layer as a function of tilt angle. Compare Figures 1.5(a) and (b)**

| Increase in layer width | Tilt angle |
|---|---|
| 25:1 | 2° 20′ |
| 20:1 | 2° 50′ |
| 15:1 | 3° 50′ |
| 10:1 | 5° 40′ |
| 5:1 | 11° 30′ |
| 2:1 | 30° |
| 1.5:1 | 41° 50′ |

They are plastic materials, to which metal powders such as iron, silver, copper, or graphite have been added. Another excellent technique is to embed the sample in low-melting bismuth alloys; in addition to electrical conductivity, a good adhesion is achieved due to the expansion of bismuth during the solidification. (This technique is rarely used because of its handling difficulty.) The following alloy compositions (in wt%) and their corresponding melting points are suitable:

| | | |
|---|---|---|
| Wood's metal | Bi50-Pb25-Sn12.5-Cd12.5 | 65 °C |
| Rose's alloy | Bi50-Pb25-Sn25 | 94 °C |
| Lipowitz's alloy | Bi50-Pb27-Sn13-Cd10 | 70 °C |
| Newton's alloy | Bi53-Sn26-Cd21 | 103 °C |

**Infiltration.** Specimens with voids, pores, and small cracks, and also sample material that is fragile or brittle are prone to breakouts, in which abraded particles generated during grinding and polishing can be entrapped. Therefore, special infiltration techniques can be used in which the mounting material fills all voids and forms a strong bond with the material. The well-dried samples are placed in their molds into an infiltrating apparatus (a desiccator with a waterjet pump or a commercially available equipment) and are evacuated for approximately 10 min. While under vacuum, the mounting material is then cast over the sample; air is admitted to the chamber, and the mounting medium is pushed into all cavities that are connected to the sample surface. The infiltrating material, an epoxy resin, must be adjusted with 10% more hardener (catalyst) than for regular mounting, since it evaporates faster at lower pressure. This technique also eliminates air bubbles that were introduced during the mixing process.

The infiltrated samples cure in approximately 16 h and are then prepared as usual. The infiltrated compound appears gray under the microscope. To improve the contrast, a red dye—Safranin—or a fluorescent powder that can be easily distinguished can be added to the resin. When using fluorescence, a mercury vapor lamp and a special set of filters are necessary.

**Mounting of Powders.** Metal powders with a particle size above 2 μm should be cold mounted in epoxy resin with pressurizing equipment. Powders that tend to agglomerate with a particle size less than 2 μm (e.g., ceramic powders) can be embedded as follows: the powder is placed in a test tube and put under vacuum with a three-way tap using running water; it is then infiltrated with a mixture of methacrylic acid methyl ester and 1% (wt%) of benzole peroxide. This mixture cures in 12 h at 50 °C, with the test tube kept closed. The test tube is then broken and the powder sample is processed as usual.

## 1.4 Identification (Marking)

Specimens are useful only if their origin is known. Therefore, specimens should be clearly marked so that their identity is not lost at any time in the preparation sequence. It is important that marking does not cause any changes to the surface of the specimen. Methods such as stamping or vibration engraving produce local deformation. When electroengraving, which utilizes the principle of spark evaporation, local changes in composition can take place through reaction with the surrounding atmosphere. These changes can alter the microstructure, through decarburization or remelting and quench cooling. Heat treated specimens are particularly susceptible to such effects. Markings should be placed on the side opposite the critical face to avoid damage. Identification labels should be fixed on the mounts immediately to avoid loss of identity at this point. Figure 1.6 summarizes the different types of identifying methods.

## 1.5 Grinding and Polishing

For microscopic examination, a sample has to be planed and smoothed in such a way that the surface becomes highly reflective, free of scratches and deformation. *Planing* includes abrasive methods, such as mechanical grinding, lapping, microtoming, fine milling, fine turning, and filing. *Smoothing* includes mechanical polishing, chemical polishing, electrolytic polishing, and combinations of these techniques whereby the mechanisms during the removal of the material are different.

Coarse and fine grinding are the initial steps in producing a satisfactory polished specimen. Together with polishing, grinding serves to smooth the surface, remove the altered surface material, and reveal the true structure of the specimen. These processes are carried out in several steps from coarse to fine so that the deformed layer can be effectively reduced. Coarse grinding removes heavy surface damage and produces a flat surface comparable to filing or planing, which are

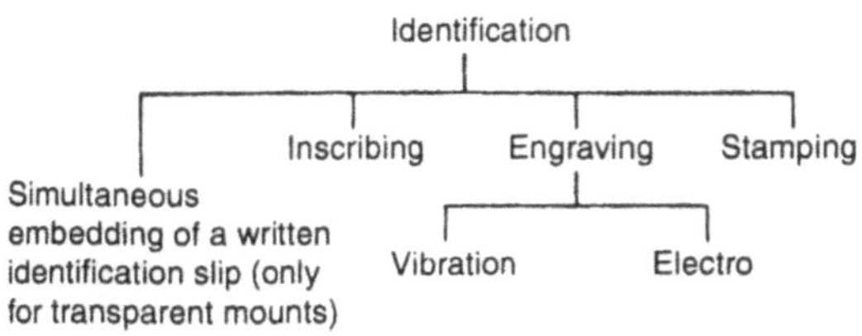


**Fig. 1.6** Methods for identifying metallographic specimens

seldom used in metallography. The surface roughness resulting from coarse grinding is in the range of 10 to 100 μm (1 μm = $10^{-3}$ mm = 3.9 × $10^{-5}$ in.). Grinding leaves a roughness of 1 to 10 μm, and fine grinding, about 1 μm. Although grinding produces a flat, smooth surface, it is suitable for macroscopic examination and macroetching only.

Because there is no sharp line between grinding and polishing, there is a considerable overlap between fine grinding and coarse polishing. For this reason, grinding and polishing are described together here. Figure 1.7 summarizes the numerous techniques presently in use.

### *1.5.1 Mechanical Grinding and Polishing*

Mechanical grinding and polishing methods are most common. They are usually accomplished by abrasive removal, using a manual technique on fixed abrasives, vibrating platens, rotating wheels, or continuous belts.

> During mechanical grinding and polishing, the surface is removed by the abrasive in the form of shavings or chips. To a lesser degree, removal takes place through adhesive interaction between the sample and polishing support. Scratches, deformation, and some smearing are characteristic consequences of mechanical grinding and polishing.

The abrasive particle size decreases from one step to the next during grinding and polishing. The coarser particles are firmly bonded to their substrate surface, while the finer particles are added loosely as a slurry or paste onto the substrate.

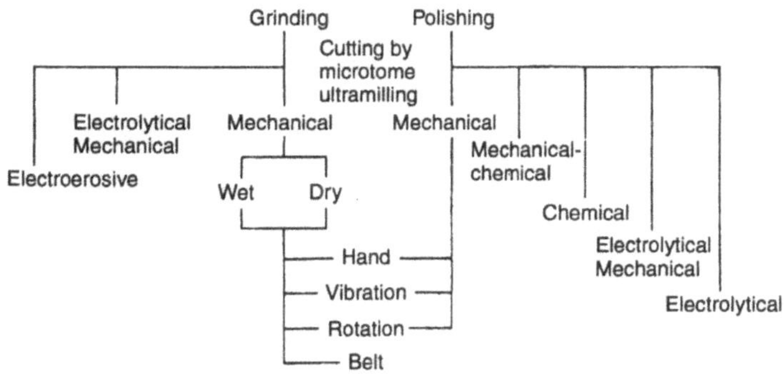


**Fig. 1.7** Methods of metallographic grinding and polishing

Working with fixed particles is considered grinding, and working with loose particles is considered polishing.

The purpose of grinding and polishing is to produce a suitable surface for microstructural evaluation by means of a step-by-step removal of deformed material, until the deformation previously introduced becomes negligible.

The depth of the scratches (roughness) with the depth of the surface deformation amounts to the total depth of the disturbed surface layer, as illustrated in Fig. 1.8; this figure is based on references from the literature, using steel as an example. The depth of the surface roughness is directly proportional to the abrasive particle size; however, the depth of the surface deformation approaches a constant value after the initial increase. The curve representing the total depth of disturbed material is the sum of both curves, depth of the surface roughness, and depth of the surface deformation.

A good understanding of the relationship as illustrated in Fig. 1.8 is very important for the successful application of mechanical grinding and polishing. In practice, the curves will be actually followed "backward," that is, in the direction of decreasing abrasive particle size. Removal of the disturbed surface layer is the key to successful optical measurements and for microhardness measurements.

> It is not adequate to remove roughness (visible scratches) from the surface without removing the deformed layer as well. Coarse grinding should be kept to a minimum because there is a tendency to produce severe deformation. Since depth of deformation decreases sharply after a grit size of about 500, increased time should be spent in removing previous damage. Care

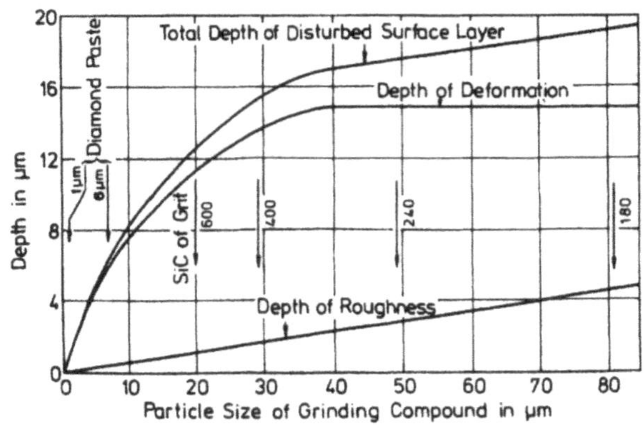


**Fig. 1.8** Depth of surface roughness and deformation and total depth of the disturbed surface layer as a function of the abrasive particle size

must be taken to prevent cross contamination of coarse abrasive particles in the finer steps. (Careful cleaning between steps is essential!)

While deformation is predominant during coarse grinding, smearing due to plastic flow is more prevalent during polishing. This effect is illustrated schematically in Fig. 1.9, in which it is assumed that polishing produces a perfectly flat surface. The troughs of the scratches have been filled with material previously protruding above the surface. This is the result of the rolling action of loose, unsupported abrasive particles during the polishing steps. This filling effect produces the smeared layer, which is markedly black in Fig. 1.9. These regions consist of a mixture of severely deformed sample material and polishing medium. Adjacent to this layer is a deformation zone perpendicular to the surface, decreasing continuously toward the interior of the sample. The lines of the equivalent degree of deformation do not run parallel to the surface, but follow the surface profile that had been generated by the coarsest grinding step.

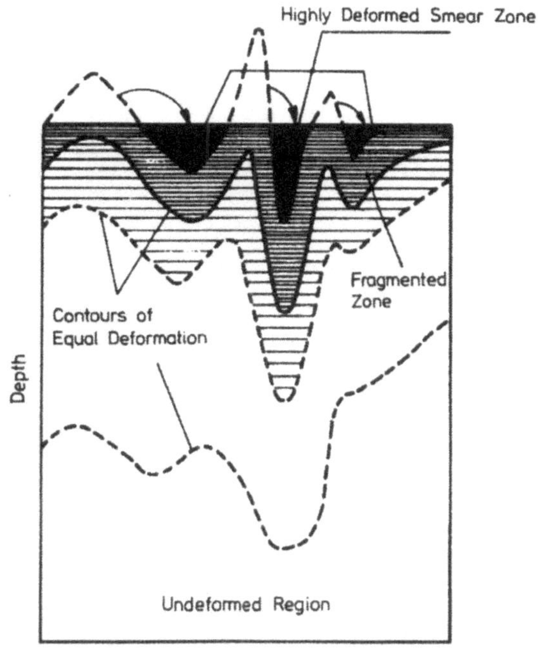


**Fig. 1.9** Schematic showing the cross section of a surface region from a mechanically ground and polished sample (sectioned 90° to the surface)

These different material states at the surface have a significant effect on many properties, for example, the electrochemical potentials. Areas with the highest degree of deformation may be more readily attacked by etchants. This causes the reappearance of scratches on a polished sample after etching, which is usually an indication of inadequate sample polishing. A large number of parameters relating to the mechanical grinding and polishing processes have to be considered; they are described in this section.

**Grinding and Polishing Compounds.** Inorganic powders are used as abrasives in metallography. The most important ones are listed with their applications in Table 1.5. Their sequence indicates that a given material is able to scratch the flat surface of the one above in the scale and is therefore harder.

Grinding and polishing compounds are classified by the supplier and should essentially have a uniform grain size. Different designations exist; for example, the grit size indicates the mesh number per linear inch for a standard sieve through which the largest particles of a given fraction will still pass through. The limit against the smaller particle size is determined by the mesh number of a sieve for the next grit size. Table 1.6 shows the commercial grit size numbers for SiC wet-grinding papers according to FEPA standards (Fédération Européenne

**Table 1.5 Grinding and polishing compounds and their application**

| Grinding and polishing compounds, listed with increasing hardness | Comments |
|---|---|
| Pumice (mixtures of $SiO_2$ and $Al_2O_3$) | Mild abrasive compound. Occasionally used for silver |
| Kieselguhr and Tripoli (mixtures of $SiO_2$ and $Al_2O_3$) | Rough polishing for brass and precious metals |
| Tin dioxide ($SnO_2$, tin stone) | Final polishing of ores |
| Magnesia (MgO, Magnesia usta) | Polishing of Mg, Al, and their alloys |
| Cerium oxide ($Ce_2O_3$) | Rarely used for metallography, mainly for glass |
| Iron oxide ($Fe_2O_3$, red polish) | Polishing of soft metals and alloys |
| Quartz ($SiO_2$, sand) | Final polishing for soft metals |
| Chromic oxide ($Cr_2O_3$, green polish) | Polishing of hard metals and alloys, e.g., Cr, steels |
| Emery (50–70% $Al_2O_3$ remainder: $SiO_2$, $Fe_2O_3$, and $Fe_3O_4$) | Rarely used as grinding and polishing compound |
| Natural corundum (90–95% $Al_2O_3$) | Abrasive wheels for coarse grinding of steels, brass, etc. |
| Synthetic corundum or electrocorundum (60–95% $Al_2O_3$; remainder: $SiO_2$, $TiO_2$, $Fe_2O_3$. Variations: precious corundum, alundum, and corundum with $Al_2O_3$ > 99%) | Important abrasive material for steel, steel castings, sintered metals, and hard steels |
| Alumina ($Al_2O_3$; modifications: cubic gamma aluminum oxide and hexagonal alpha aluminum oxide) | Commonly used polishing compound. Suitable for nearly all metallographic preparation |
| Silicon carbide (SiC, carborundum) | Commonly used grinding material. Suitable for nearly all metallographic preparation |

des Fabricants de Produits Abrasifs) and according to the ANSI (American National Standard Institute) standard with their actual particle size in μm. The information of the particle size distribution is very important since they also include the largest particles that will lead to the deepest scratches. Most suppliers list only the average particle size, but this should only serve as a guideline for the selection of grinding papers or grit numbers. However, knowing the particle size distribution is very important, because a tight particle size distribution translates into a smaller deviation from the average particle size.

> Sharp edges, high hardness, high particle density, and rigid bonding to the support material substantially increase the material removal and reduce the depth of deformation and smeared layer. Diamond is superior to the other abrasives, for example alumina, because of the extent to which it meets all the above requirements.

**Table 1.6 Grain size designation for SiC wet-grinding papers**

| FEPA standard grit size | ANSI standard grit size | Mean particle size, μm | Particle size distribution, μm |
|---|---|---|---|
| ... | 60 | 325 | ... |
| 60 | 80 | 260 | 500–212 |
| 80 | ... | 196 | 355–150 |
| 100 | 100 | 154 | 300–125 |
| ... | 120 | 137 | ... |
| 120 | ... | 120 | 212–90 |
| ... | 150 | 110 | ... |
| 150 | 180 | 95 | 180–75 |
| 180 | ... | 75 | 150–63 |
| ... | 240 | 71 | ... |
| 220 | 280 | 65 | 125–53 |
| 240 | ... | 59 | 110–44 |
| 280 | 320 | 52 | 101–39 |
| 320 | ... | 46 | 94–34 |
| 400 | ... | 35 | 81–25 |
| 500 | 400 | 31 | 77–21 |
| 600 | 3 | 26 | 72–18 |
| 800 | 500 | 22 | 67–15 |
| 1000 | 600 | 18 | 63–12 |
| 1200 | Micro cut | 15 | 58–10 |
| 2400 | ... | 10 | ... |
| 4000 | ... | 5 | ... |

*Silicon carbide* (SiC) is the most commonly used abrasive in metallography. Its sharp-edged cutting grains tend to fracture when under pressure and thus produce fresh cutting surfaces. The particles are bonded to water-resistant papers or cloths. Their application ranges from tough to hard and brittle materials with a Vickers hardness of approximately 1200 HV.

*Diamond* is the hardest cutting material. Its particles have a crystalline sharp-edged shape and a very high cutting efficiency that prevents smearing of the specimen surface. The diamond particles can be bonded to metal, plastic, or ceramic disks used for grinding hardmetals, cermets, advanced ceramics, and composites. Loose particles are available for lapping and polishing with particle sizes ranging from 45 to 0.25 μm. They are sold as paste, spray, or suspension that are applied to the polishing disk. Additional lubrication is necessary to provide an optimum distribution of the diamond particles and to reduce their wear.

*Alumina* ($Al_2O_3$) is an economical and overall suitable polishing compound. Its use for planing and smoothing depends not only on the particle shape and particle size distribution, but also on the proportion of the two-crystal modifications gamma and alpha, with their tendency to agglomerate. Alumina is available as powder, pastes, or water suspensions in particle sizes of 5, 1, 0.3, and 0.05 μm.

*Magnesia* (MgO) is preferred for final polishing of soft metals such as aluminum and copper. A slurry with hot, preferably boiled water (free of carbonates) is prepared and then filtered through a cloth. Only small amounts should be prepared since magnesia reacts with air to form magnesium carbonate.

*Silicon dioxide* ($SiO_2$, quartz, silica) in an alkaline suspension is a polishing compound, and its use has been increasing over the past years as a final polishing material. Even very difficult and very ductile materials can be polished successfully to a scratch-free surface. Yet its application is not entirely without problems; it crystallizes rapidly in air, and skin should not be exposed to the slurry. When using it with an automatic polishing equipment, it is recommended to have the slurry in a tray.

**Grinding and Polishing Substrates.** Grinding and polishing compounds together with their substrate form one entity. Both have to be compatible. Table 1.7 shows a survey of common grinding and polishing substrates (support materials). They are listed according to their individual "hardness" and should be chosen according to their application.

*Substrates for grinding* are ceramic grinding materials (e.g., corundum stones), papers or textiles on which the grinding particles are incorporated, or plastic and metal disks. Most common are special papers or textiles, onto which the particles are glued.

*Substrates for polishing* are generally textile materials (felt, velvet, synthetic materials, linen, wool, etc.), and they are offered by the supplier for the use of varied and specific polishing methods. Since the particle size for polishing is much finer, it is not possible to solidly fix the particles onto the substrate as is done for the grinding materials. Polishing compounds are therefore added as slurries onto the substrate or in the form of paste. The polishing substrates must be such that they can retain the loose polishing compounds and offer some light adhesion. They should therefore have a structured surface, good abrasion resistance, must retain their shape, and should be dense and resistant to the polishing compound. The metallographic supplier usually identifies and labels polishing cloths by their material and fabrication type. The recipes listed in this book refer to some of these common tradenames.

It is nevertheless still important to evaluate and select the appropriate disk or cloth for a certain material application from the large number of commercially available goods. Table 1.8 shows a list of available polishing cloths.

The following rules apply in principle:

- The depth of surface roughness and deformation decreases with increasing elasticity of the support material.
- Rigid support materials produce flat surfaces, but tend to scratch the surfaces of softer materials.
- Soft support materials produce well-polished surfaces, but can promote edge rounding and relief effects.

**Lubricants for Grinding and Polishing: The Surrounding Media.** Grinding can be carried out wet or dry. Normally, the specimens are ground with running water on SiC paper or diamond disks. The water acts as a coolant to the sample surface and removes the abrasion debris. Wet grinding produces a slightly higher

**Table 1.7 Common grinding and polishing substrates**

| Sample material | Grinding | Lapping | Polishing(a) |
|---|---|---|---|
| Hard | Corundum disks<br>Diamond disks (plastic bonded)<br>Steel net cloth | Cast metal disk<br>Composite disk of metal/plastic | Perforated chemical fiber cloth (Pellon)<br>Hard chemical fiber cloth (Pellon)<br>Broadcloth, synthetic |
| Composites, cermets | Diamond disk (plastic bonded)<br>SiC paper | Composite disk of metal/plastic | Hard silk cloth, woven<br>Cotton cloth, woven<br>Felt cloth<br>Short nap cloth of synthetic fiber, silk or velvet |
| Medium | SiC paper | Only where edge retention is required | Long nap cloth of synthetic fiber, silk, or velvet |
| Soft | SiC paper | ... | Synthetic cloth chemical resistant |

(a) Select polishing cloths according to their specific application.

deformation depth than dry grinding. The explanation is the so-called Rehbinder effect, according to which the metallic materials appear softer the higher the surface tension of the surrounding medium is, and in effect the surface tension of water is relatively high. This effect can be suppressed by adding detergents. The deformation depth of soft materials can also be reduced by dry grinding on the finer grinding papers treated with some wax or stearin. Samples that have a high corrosion sensitivity can be ground with a low-viscosity oil as a lubricant.

For diamond pastes and sprays, two types of lubricants are used: a low-viscosity alcohol-based lubricant for harder materials and a high-viscosity oil-based lubricant for softer materials. Both are water and alcohol soluble.

**Table 1.8 Polishing cloths**

| Material | Application |
|---|---|
| Metal mesh disk(a) | Used for planar grinding and coarse grinding with diamond abrasives (45–15 μm) |
| Heavy napless chemotextile pad(a) | Used mainly for hard materials with diamond abrasives (15–1 μm) |
| Hard woven napless polyester pad(a) | Used for most materials after planar grinding with diamond abrasives (15–6 μm) |
| Woven napless silk PSA cloth(a) | Used mainly with harder materials, ores, minerals, and coal (abrasives ≤6 μm) |
| Napless nylon cloth(a) | Used with most materials (abrasives ≤9 μm) |
| Pure steam-processed silk (125 threads/in., non-PSA cloth)(a) | Same as woven napless silk PSA cloth |
| Synthetic woven cloth with TD weave(a) | Used with most materials (abrasives 9–1 μm) |
| Napless chemotextile pad(a) | Used for most materials (abrasives ≤9 μm). Excellent for inclusion retention |
| 100% virgin wool cloth | Used for steels (abrasives ≤6 μm) |
| Woven wool cloth | Used mainly for steels (abrasives ≤3 μm) |
| Napless canvas | Used for rough polishing with coarse diamond, silicon carbide, or alumina powders |
| Napless cotton cloth | Used for softer materials (abrasives ≤3 μm) |
| Synthetic velvet short-nap cloth | Used for soft materials with alumina and colloidal silica abrasives (≤3 μm) |
| Synthetic medium-nap rayon cloth | Used for most materials (abrasives ≤6 μm) |
| Synthetic, chemically resistant low-nap cloth(a) | Used for most materials for final polishing (≤1 μm) with alumina and colloidal silica abrasives |
| Medium-nap cotton cloth | Used for most materials, particularly softer materials for final polishing (≤1 μm) with oxide slurries |
| High-nap synthetic velvet cloth | Used mainly for softer materials for final polishing (abrasives ≤1 μm) |

PSA, pressure-sensitive adhesive; TD, special textile weave pattern. (a) Cloths that provide the best edge retention and specimen flatness

**Pressure, Time, and Rotating Speed.** Although increasing these parameters has an advantageous removal effect, the effect on the deformation depth and smearing varies:

- Excessive pressure may cause overheating or deformation of the sample surface and can result in a change to the microstructure. Pressure should not be too high, especially during polishing.
- Short grinding times and long polishing times are preferable (Fig. 1.8)! However, excessively long polishing (overpolishing) can cause relief due to the different removal rate of microstructural constituent.
- The rotating speed of the wheels is usually between 150 and 300 rpm. A slower speed is recommended for brittle, easily deformed materials or for materials tending to recrystallize. High speeds are preferred for very hard materials (ceramics, intermetallics, hardmetals).
- The surface temperature increases with increasing grinding and polishing speed. Temperature-sensitive samples must therefore be slowly ground and polished.

**Specimen Motion During Grinding and Polishing.** The grinding and polishing motion affects the edge retention of the sample. To obtain a high degree of edge retention, the specimen surface must be maintained flat and parallel to the substrate during all grinding and polishing motions.

For an improved removal rate and to control the grinding and polishing grooves, the specimen must be rotated 90° after each grinding and polishing step. This allows examination to determine whether the grooves from the previous step have been removed. During the final polishing step, the specimen is rotated against the direction of the wheel or rotated around its own axis.

**Standard Recipe.** The standard preparation for mechanical grinding and polishing should meet the following conditions:

- Is versatile in its application
- Produces an adequate to good sample quality
- Is usable for manual or automatic preparation
- Involves reasonable time and equipment expense
- Is easy to understand
- Is expendable and variable for difficult or more specific applications

Such a standard preparation is:

- Wet grinding on SiC paper, 320/400/500/600 grit according to the ANSI Standard System, or 320/500/800/1000 according to the FEPA Standard System

- Rough polishing with diamond pastes, suspensions, or sprays, 6, 3, or 1 μm on synthetic silk or cotton cloth with an alcohol-based lubricant
- Final polishing with alumina slurry on a wool or velvet cloth
- Careful cleaning between the individual grinding and polishing steps is necessary to prevent cross contamination

Detailed methods for the preparation of metals, alloys, ceramics, and polymers are described in section 1.9 in this chapter and in sections 2.8 (in Chapter 2 "Ceramography") and 3.1 (in Chapter 3 "Plastography").

**Automatic Grinding and Polishing.** Although manual mechanical grinding and polishing is still the basis of metallographic sample preparation, automation of the individual steps is becoming more and more dominant since it offers distinct advantages:

- Time saving for the metallographer
- Efficiency with higher sample volume
- Reproducibility by defined and measurable working parameters
- Often a better sample quality regarding flatness and edge retention

However, since the samples need still to be handled for cleaning after each step and then placed manually onto the next disk, it is still only a semiautomatic operation. Some attempts for a fully automatic sample preparation are promising, but are rare since an extensive equipment expense is involved with the difficulty of adjusting to the variety of different methods.

There are three different types of equipment. Most widely used are *rotating wheels* with various systems of sample holders, that may be rotating separately or are moved by the rotating polishing wheel. These machines can be programmed for speed, pressure, polishing compound dispensing, and for time. Another type of equipment moves a single specimen with a *rotating arm across a fixed disk* that is placed in a bowl. This is used for rough and final polishing with diamond grits of 15 to 0.25 μm, as well as final polishing in polishing solutions with or without reagents. A third possibility of a semiautomatic preparation is by *vibratory polishing*. This apparatus consists of a vibrating platen that is mounted on a base with laminated springs. An electromagnet is used to excite vertical and rotational oscillations. The polishing disk is a removable bowl in which several specimens rotate along the disk as well as around their own axis. A flat, scratch-free, and deformation-free sample surface can be obtained for a range of widely different materials using combinations of a variety of polishing materials. Vibratory polishing is not suitable as a grinding operation.

**Lapping** is a special method of material removal that can be placed between grinding and polishing. Lapping is particularly important for the preparation of hardmetals, composites, and ceramics. Lapping produces excellent flatness and edge retention of the sample, maintaining low roughness and disturbed material depth. The mechanism is very similar to mechanical polishing. The lapping compound is added onto the lapping disk together with the lubricant. The loose particles roll, press, and slide on the surface, but a portion remains on the lapping disk and causes abrasive removal of the material. With the applied pressure and the relative movement between the disk and the specimen surface, the flatness of the disk is transferred to the specimen itself during the material removal at the specimen surface. As a result, the specimens maintain an excellent flatness.

Lapping disks are made of glass, plastic, wood, and metal. These are mainly used for industrial surface treatments or for the preparation of mineral specimens. Special lapping disks were developed for metallographic, ceramographic, and plastographic sample preparation; they are made of plastic and metal powder composite materials of different hardness and design. They may have a smooth disk surface, or they can contain grooves to improve the removal rate and prevent edge rounding by aquaplaning. Diamond particles together with low-viscosity lubricants are more suitable than silicon carbide, boron carbide, and corundum particles. An important criterion for the effectiveness of lapping is the applied pressure. High pressure produces more quasi-fixed particles and cutting action and increases the material removal, low pressure leaves more particles rolling and results in a lower material removal rate. Less material is removed during lapping than during grinding, but more than during polishing. Preparation time of lapping is relatively long. Semiautomatic lapping equipment is commercially available.

### 1.5.2 Microtome Cutting and Ultramilling

Both techniques require rather sophisticated commercially available equipment. They are excellent for the preparation of soft materials, composites, materials with platings, and combinations of different materials. As during grinding, the material is removed by abrasion, and the cutting knife can be considered as one large abrasive grain. The knives are made of hardmetal or diamond with adjustable cutting and clearance angles. With correct adjustment, the sample surface shows no deformation and smearing and can be examined under the microscope without any additional preparation. This method, however, is limited for materials with a Vickers hardness of up to 150. Ferrous or refractory materials are difficult to cut with diamond knives because of the strong chemical affinity (formation of carbides) that causes tearouts from the surface of the material.

During *microtome cutting,* the sample surface is planed by the cutting knife, and its final width is the width of the cutting knife.

During *ultramilling,* the sample surface is milled with a milling head at a rotation of 700 to 3200 rpm, and a larger surface area can be worked.

### 1.5.3 Electrolytic Grinding and Polishing

While electrolytic grinding and electro-erosive grinding are rarely used, and have only been mentioned to present a complete picture, electrolytic polishing is commonly used as a sample preparation technique.

> During electrolytic polishing, which is also called anodic polishing or electropolishing, the smoothing action is caused by the anodic dissolution of the sample surface in an electrolytic cell.

Figure 1.10 shows a schematic of a simple electrolytic cell that can be easily assembled. The position of the anode (sample) and the cathode can be adapted to suit the particular application. Commercially designed and produced *electrolytic polishers* are available and versatile enough to meet most requirements.

Suitable *electrolytes* for anodic polishing are usually mixtures of phosphoric, sulfuric, and perchloric acids in ionizing solutions such as water, acetic acid, or alcohol. Glycerol, butyl glycol, urea, and so forth are added to increase the viscosity. Alkaline solutions are used for metals that form highly soluble hydroxides, and cyanide solutions are used for metals that form soluble cyanides. While most of the electrolytes listed in the recipes in section 1.9 in this chapter and in section 2.8 (in Chapter 2 "Ceramography") are safe when handled according to regular safety precautions, mixtures of perchloric acid can be explosive (see sec-

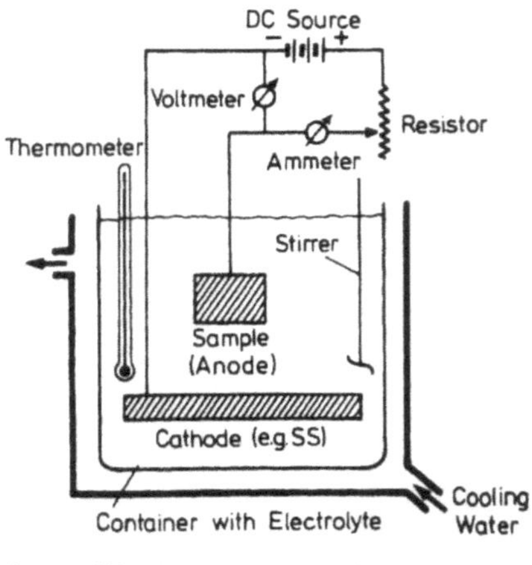


**Fig. 1.10** Electrolytic polishing cell having a series circuit

tion 4.4, "Handling Tips for Hazardous Materials" in the Appendix). The tendency of perchloric acid mixtures to explode is related to the concentration. However, caution should always be taken. Even safe mixtures can become highly dangerous, if a localized concentration increase does occur during the polishing process. The electrolyte should therefore always be well stirred and cooled during the polishing process. The polishing action for a given combination of electrolyte and sample material is controlled by the current density (A/cm$^2$) and voltage (V), the distance of the sample from the electrode, size ratio of cathode to anode, surface condition of the anode, polishing time, temperature of the electrolyte, agitation of the electrolyte, and concentration of the electrolyte. Most of these parameters are listed in the recipes of section 1.9 in this chapter and section 2.8 (in Chapter 2 "Ceramography"). They are omitted when they are not critical, if they can be easily established, or have been supplied with the equipment.

A typically ideal relationship between current density and applied voltage for many common electrolytes is shown in Fig. 1.11. The current-density/voltage curve includes four characteristic parts:

- Anodic material dissolves between A and B by releasing electrons:

$$Me \leftrightarrow Me^{2+} + 2e$$

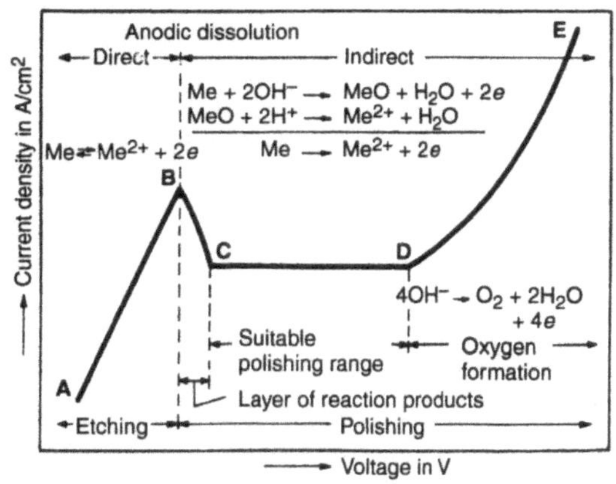


Fig. 1.11 Idealized conditions of current density versus applied voltage. See text for discussion.

A layer with an increased concentration of metal ions forms at the anode surface. The portion A–B of the curve is important for electrolytic etching (see section 1.7 in this chapter).

- At B, a potential that deposits a reaction product (e.g., an oxide) is attained, and a passive layer forms on the sample surface. As a result, the current density drops to a constant value:

$$Me + 2OH^- \leftrightarrow MeO + H_2O + 2e$$

- The electrolyte composition must be able to dissolve the passive layer:

$$MeO + 2H^+ \leftrightarrow Me^{2+} + H_2O$$

The formation and dissolution of the passive layer reaches an equilibrium through alternating diffusion and electrolytic processes that form the plateau C–D in the current-density/voltage diagram. Electrolytic polishing takes place in this range.

- Oxygen evolution begins at D:

$$4OH^- \leftrightarrow O_2 + 2H_2O + 4e$$

The initial oxygen evolution is low, and the settling time of the gas bubbles is relatively long. This interrupts the polishing process and pitting occurs. By increasing the voltage (D–E in the diagram), the oxygen evolution increases as well, and the settling time of the gas bubbles on the anode surface is then reduced.

Though *anodic polishing* takes place from point B in the diagram, it is only between C and D where other effects (formation of a passive layer or evolution of oxygen) no longer interfere. Therefore, most metallographic polishing recipes refer to the C–D portion of this diagram. The D–E portion is rarely used, and if then only for industrial-type electropolishing (anodic brightening).

The different regions of the current-density/voltage diagram are in most cases not very pronounced in the *actual* curves. To distinguish individual points is often impossible when using electrolytes with a very high resistivity.

The original surface condition of the anode (sample) affects the polishing time. Typically, the polishing time becomes shorter, the higher the current density and the better prepared the sample surface.

A prerequisite for electrolytic polishing is a sample material that is electrically conductive, for example, all metals and alloys, as well as some nonmetallic materials such as carbides and graphite.

All single-phase metals and alloys can be electrolytically polished without any problems. Multiphase alloys are less suitable for anodic polishing because of the different electrolytic potentials between their individual phases. However, electrolytical polishing methods can also be successful for some multiphase alloys.

*Shock polishing* is a method that can be applied for the final polishing of multiphase microstructures. With this technique, the voltage is increased by approximately 50%, the solution is agitated at a high rate, and the time is very short (¼ to ½ s); this prevents a selective attack in the multiphase material. If the result is not satisfactory, the process should be repeated but without an increase in time.

Listed below are the advantages of electrolytic polishing:

- No deformation and smeared surface layers
- Short time and good reproducibility; applies only when using a suitable recipe
- No or only little heat generation of the sample
- Possibility for subsequent etching with the same equipment
- Easy removal of deformation or smeared surface layer from mechanical grinding or polishing with a short electrolytic polish

However, the lack of good recipes for many multiphase materials and the disadvantages listed below present certain limitations of electrolytic polishing for this material:

- Edges are especially heavily attacked. Conductive mounting materials with their good adhesion properties may help to support the edges. However, good edge retention is very difficult to obtain.
- Only slight surface roughness is being leveled; large, uneven areas are not well polished.
- A surface layer is formed on the polished surface.
- Electrical contact can be lost with samples that oxidize easily. Special provisions have to be made for nonconductive mounting materials (see section 1.3.3 in this chapter).
- Coarse-grained materials are not well suited for electrolytic polishing.
- At the interface of nonmetallic inclusions the matrix metal is preferentially attacked, resulting in heavy relief and gap formation.

### 1.5.4 Chemical Polishing

This is a very simple technique. The sample surface is immersed into a polishing solution (electrolyte) and agitated for several seconds or minutes and then rinsed. The polishing solution dissolves the surface at a uniform rate and rapidly forms a barrier layer saturated with metal ions. The peaks of the surface are the ones exposed first to the fresh and agitated solution and dissolve first. The same action occurs at a much slower rate in the grooves of the sample surface; the material removal there is low while the surface peaks are preferentially dissolved, thus producing an overall flat and deformation-free surface.

The *polishing solutions* are usually mixtures of highly oxidizing reagents such as nitric, sulfuric, and chromic acids or hydrogen peroxide. Viscous agents are also added to control diffusion and convection rates, making the process more uniform. Phosphoric acid in particular forms a viscous film with the metal ions. Most polishing solutions are quite insensitive to concentration variations.

The advantages of chemical polishing are:

- Simplicity and economy.
- Little initial preparation is required; 320 grit finish is adequate. When cut with an abrasive cutoff wheel, the sample may be directly polished without further preparation.
- Simple post-treatment of the sample; generally, rinsing in water is adequate.
- Polished surface is free of deformation and smeared layers that may occur after mechanical polishing.
- The specimen does not need to be electrically conductive.

However, the technique has certain disadvantages:

- Edges are heavily attacked, causing rounding.
- Coarse-grain materials show an orange-peel effect.
- Only the small-scale surface roughness is smoothed; larger grooves or peaks are rounded but not flattened.
- A surface film composed of reaction products forms.

### 1.5.5 Combined Polishing Methods

A combination of mechanical polishing with chemical or electrolytic polishing methods can often produce better results than using only one technique.

**Multiple Polishing.** When electrolytic polishing is used with heterogeneous alloys, some undesirable surface effects such as relief formation or surface layers may be produced. A brief intermediate mechanical polishing step may remove this condition and produce a satisfactory finish.

**Attack polishing** consists of using an intermediate application of a chemical etchant during the final polishing sequence; it is especially useful in avoiding flow layers on soft metals and alloys, as well as refractory metals.

**Etch polishing or chemical-mechanical polishing** is a combination of mechanical material removal with a simultaneous chemical attack; highly diluted etchants for the material to be polished are used. This technique is especially useful for:

- Soft and ductile materials, to promote the removal of deformed and smeared layers
- Materials with high abrasion resistance (e.g., refractory materials)
- Multiphase alloys
- Coated materials; it is possible to produce flat and scratch-free microstructures of combinations of different materials, for example, platings by a sequence of etch polishing steps and by alternating the etchant that will attack the particular phases present respectively.

A sample surface finish of 15 μm is usually satisfactory. The polishing slurry is alumina with water and an appropriate chemical reagent. Suitable polishing equipment are vibratory polishers or disks with a retaining rim (polishing bowls) and an automatic sample holder attachment.

**Electrolytic lapping** is the simultaneous electrolytic and mechanical removal of material. In this technique, the specimen (anode) is placed on a cloth saturated with electrolyte and mounted on a support or wheel that serves as the cathode. A schematic shows the combined anodic and mechanical polishing on rotating disks in Fig. 1.12.

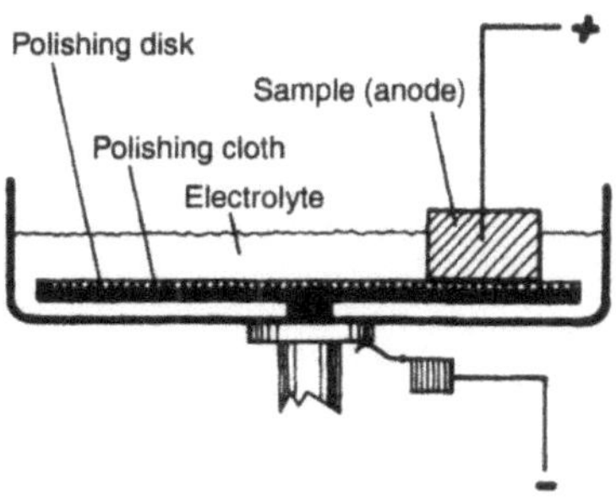


**Fig. 1.12** Electrolytic lapping technique

Although direct current is most frequently used for electrolytic lapping, an alternating current at low frequency produces superior results with some special materials like molybdenum, tungsten, and rhenium. Dilute solutions of sodium thiosulfate, nitric acid, oxalic acid, picric acid, hydrogen peroxide, and so forth are suitable for both electrolytic lapping and etching. The addition of abrasives such as alumina has no effect in most applications.

> Electrolytic lapping combines the advantages but also some of the disadvantages of the electrochemical and mechanical polishing. It is well suited for homogeneous as well as heterogeneous metals and alloys. As a complementary technique, electrolytic lapping has been proven successful when the electrolytic or the mechanical polishing alone fails.

## 1.5.6 Evaluation of Polishing Methods

Reflectivity of the finished surface is a sensitive criterion for judging quality. Deviations from the reflectivity of ideal surfaces with an atomic roughness give an indication of the effect of different polishing methods; for example, cleavage surfaces of crystals or surfaces of single crystals grown under an ultrahigh vacuum are good reference standards. Therefore, reflectivity is suitable for evaluating the effectiveness of individual polishing methods.

Figure 1.13 shows the mean deviation, in percentage of reflectivity, of polished surfaces from the ideal surfaces of various materials. The baseline represents reflectivity values for freshly cleaved surfaces—the ideal values of surface quality obtainable. Mean deviations for different polishing methods are plotted for different materials. Obviously, any polishing treatment will produce a lower quality than the ideal cleaved surface. Of the popular polishing techniques, alumina polishing produces the poorest values, while electrolytic and diamond-knife microtoming produce results approaching the ideal values.

It is also obvious from Fig. 1.13 that the effectiveness of different polishing methods varies from material to material. Antimony trioxide ($Sb_2O_3$), for example, shows the same deviation when prepared with alumina slurries as gold and copper do after electrolytic polishing. Properties other than reflectivity are also affected by the polishing technique. Severe grinding and polishing stresses have a noticeable effect on microhardness. When reporting surface condition, information concerning the grinding and polishing techniques should be included. Differences in actual results compared with expected results may stem from differences in the manner of preparation. For routine preparation and examination of microsections, these considerations are of secondary importance. Alumina

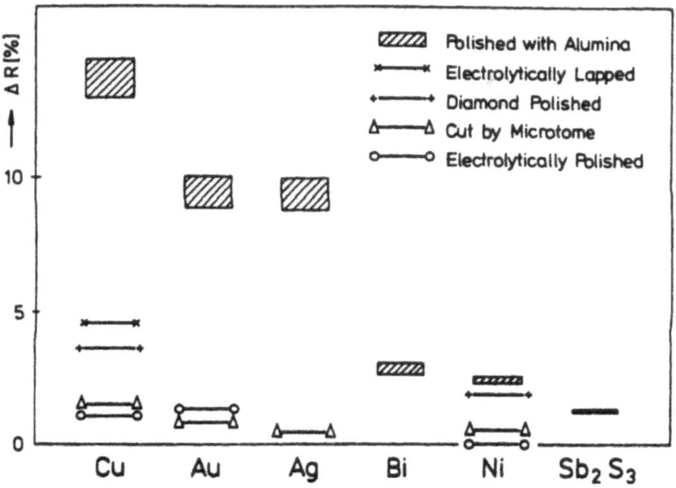


**Fig. 1.13** Mean deviations, $\Delta R$, in percent reflectivity, of a polished surface from an ideal surface

and other oxides are widely used because they produce acceptable results at a reasonable cost.

## 1.6 Cleaning

Cleanliness is an important requirement for successful sample preparation.

> All remnant grinding and polishing compounds have to be removed completely from the sample after each step to avoid cross-contamination, which would reduce the efficiency of the next preparation step. It is often even necessary to clean the sample several times within one preparation sequence. This is done by rinsing the sample under running water and by additional wiping with cotton or by ultrasonic cleaning; other methods can be used in specific cases, as shown in Fig. 1.14.

*Ultrasonic cleaning* is the most effective cleaning technique. Remnant polishing compounds can be completely removed, even from very fine cracks, pores of porous or fractured samples, or at the interfaces between the specimen and the mounting material. Ultrasonic cleaning removes even very persistent dirt residues. Usually ultrasonic cleaning needs only 10 to 30 s.

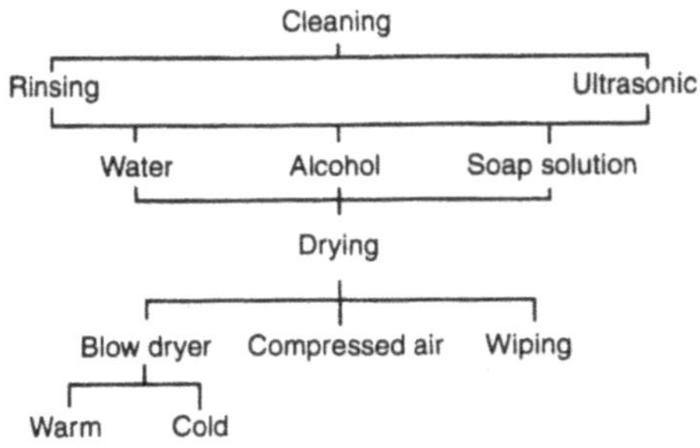


**Fig. 1.14** Cleaning methods for metallographic samples

To *dry* the specimen surface, the sample is rinsed with either alcohol, benzene, or ether. With their low boiling points, these solvents dry quickly and do not stain when using a hot air dryer. Any remaining liquid in cracks and pores has to be evaporated. A different method is to dry the sample surface with cleaned compressed air.

A clean surface, free of oil, is a prerequisite for chemical or electrolytic treatment especially for etching, since any traces of dirt can affect the overall evaluation.

## 1.7 Metallographic Etching

The final polished sample surface should be examined first under the microscope with brightfield illumination; features, such as pits, pores, cracks, inclusions, contamination, changes caused by corrosion attack, and so forth can clearly be distinguished.

In most cases, a polished specimen will not exhibit its microstructure because incident light is uniformly reflected. Since small differences in reflectivity cannot be recognized by the human eye, some means of producing image contrast must be employed. Although this has become known as "etching" in metallography, it does not always refer to selective chemical dissolution of various structural features.

There are numerous ways of achieving contrast. These methods may be classified as "optical," "electrochemical (chemical)," or "physical," depending on whether the process alters the surface or leaves it intact (Fig. 1.15).

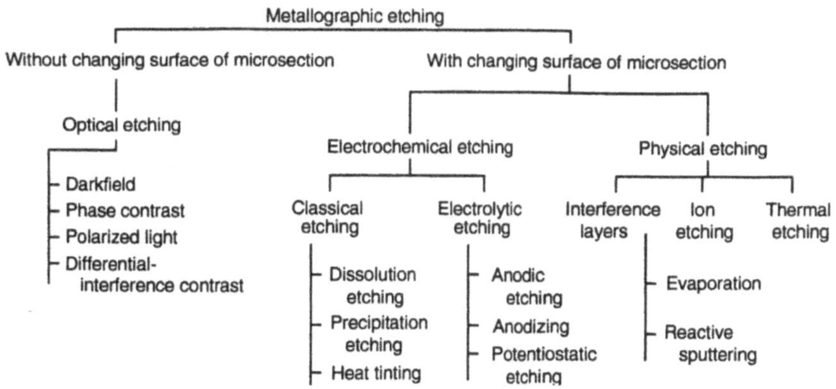


**Fig. 1.15** Methods of etching

## *1.7.1 Optical Etching*

"Optical etching" uses special illuminating techniques, all of which are based on Köhler's illumination principle that is used for general brightfield illumination. Brightfield illumination is characterized by normal reflection of the incident rays from the sample surface through the objective lens. Details of different reflective intensity appear in different shades of dark or light; this difference must be at least 10% to be noticed by the human eye. When this is not the case, additional illuminating techniques can be used for different types of examination.

In *darkfield illumination,* only the light rays deflected by diffuse scattering of surface roughness or semi-opaque phases are collected by the objective lens. This technique permits easy differentiation of pores and inclusions, very fine crack propagation, and easy identification of semi-opaque phases by their inherent color (e.g., copper (I) oxide as purple red).

*Phase contrast* is mainly used for transmitted light microscopy. Small differences of the density and the refractive index of individual constituents can be revealed by transforming phase displacements of direct and diffracted light into detectable variations of brightness intensities. For incident light microscopy, phase contrast is usually replaced by the differential interference contrast technique.

*Interference contrast* takes place between two beams of light, the object and the reference beam. Slight differences in height can be revealed with this technique. The best known is the differential interference contrast (DIC) technique by Nomarski.

*Polarized light* is used to differentiate between optical anisotropic phases and optical isotropic phases. Optical anisotropic, or noncubic phases, exhibit their characteristic colors with four intensity peaks each, when examined in white linear polarized light by rotating 360°. Many intermetallic compounds and slag inclusions can be identified with polarized light by their characteristic anisotropic effects.

Domain structures of ferri- and ferromagnetic phases can be easily revealed because of the magneto-optical *Kerr effect*. For this technique, the oscillating plane of the polarized light is slightly rotated by the reflection of a domain, which depends on the direction and the amount of magnetism; the magnetic domains will be apparent in different shades. Magnetic phases can be distinguished by their characteristic domain structures using this technique.

The accessories needed for the above-described techniques are normally available for most commercial microscopes. Considering the relatively simple equipment manipulation and the good accuracy obtained, it is unfortunate that these different illumination techniques are not as fully exploited as one would assume. The polished surface does not have to undergo any treatment with these techniques. They save time and material and eliminate the possibility of misinterpretation of microstructural artifacts.

## *1.7.2 Electrochemical (Chemical) Etching*

During all electrochemical etching of metallic specimens, which is also referred to as "chemical etching," reduction or cathodic reactions and oxidation or anodic reactions take place (redox process). Reduction is the absorption of electrons (cathodic reaction); oxidation is the emission of electrons (anodic reaction):

$$\text{oxidizing agent} + \text{electrons} \leftrightarrow \text{reducing agent}$$

$$Fe^{3+} + e \leftrightarrow Fe^{2+}$$

Electrolytic grinding and polishing (section 1.5.3), chemical polishing (section 1.5.4), and several combined polishing methods (section 1.5.5) are based on redox processes, that is, electrochemical reactions. This is not always obvious from conventional terminology.

All metals contacting the etching solutions have a pronounced tendency to become ionized by releasing electrons. The extent of this reaction can be recorded by measuring the electrochemical potential and by comparing the potential of the metal to a reference electrode. (Hydrogen electrode: a platinum sheet is surrounded by hydrogen at 1 atm pressure and immersed into an aqueous solution of a hydrogen ion activity of $a_H = 1$.) The tabulation of various metals becomes the electromotive series of elements:

$Li^+$, $Na^+$, $K^+$, $Ca^{2+}$, $Ba^{2+}$, $Be^{2+}$, $Al^{3+}$, $Mn^{2+}$, $Zn^{2+}$, $Cr^{3+}$, $Cd^{2+}$, $Tl^+$, $Co^{2+}$, $Ni^{2+}$, $Pb^{2+}$, $Fe^{3+}$, $H^+$, $Sn^{4+}$, $Sb^{3+}$, $Bi^{3+}$, $As^{3+}$, $Cu^{2+}$, $Ag^+$, $Hg^{2+}$, $Au^{3+}$, $Pt^{3+}$

The elements are listed in decreasing order of their electro-affinity. All elements preceding hydrogen are attacked by acids with an evolution of hydrogen; those elements listed after hydrogen cannot be attacked by acids without the addition of an oxidizing agent. Thus, microstructural constituents of different electrochemical potential are attacked at different rates. This produces differential etching resulting in microstructural contrast.

Electrochemical etching may be regarded as "forced corrosion." The differences in potential of the microstructural constituents create a subdivision into a network of adjacent miniature anodic and cathodic cells (local elements). These local cells originate not only from differences in the chemical composition of phases, but also from the irregularities in the crystal structure, present as grain boundaries or other inhomogeneities, such as:

- Inhomogeneities of the dislocation density (cold-worked regions are more readily attacked than undeformed material)
- Irregular formation of oxide layers (lustrous areas are less resistant)
- Concentration fluctuation of the electrolyte or the etching solution (low concentration produces a lesser attack)
- Differences in the electrolyte flow rate (faster flow rates promote attack)
- Differences in the oxygen content of the electrolyte (aerated solutions decrease attack)
- Differences in the illumination intensity also cause differences in potential

Because of the potential differences between the individual microstructural constituents, the dissolution rate of the sample surface varies as well, producing the contrast.

A wide variety of *etchants* are available, including acids, base alkaline solutions, neutral solutions, mixtures of solutions, molten salts, and gases; many examples of etching solutions are listed in section 1.9 in this chapter and sections 2.8 (in Chapter 2 "Ceramography") and 3.2 (in Chapter 3 "Plastography"). Most of these recipes were derived empirically. Their composition and mode of application can easily be varied and modified and also are useful for materials other than those mentioned in the recipes.

The *rate* of the etch attack is mainly determined by the degree of dissociation and the electrical conductivity of the etchant. Both can generally be adjusted in a certain way by adding other chemicals. This may explain why many etchants contain small amounts of other substances whose significance is not immediately apparent.

The *stability* of many etching solutions is limited, since redox potentials change with time. Changes may also occur while the etchant is in use, so that it must be discarded after a limited time.

*Etching times* range from seconds to hours. When instructions are given, progress is judged by the appearance of the surface during etching. Usually the surface becomes less reflective (duller) as etching proceeds. Etching time and *etching temperature* are closely related; by increasing the temperature, the time can usually be decreased. However, this may not be advisable because the contrast could become uneven when the rate of attack is too rapid. Most etching is performed at room temperature. Sources of error are numerous, especially in electrochemical etching. Etching errors may lead to microstructural misinterpretation. For example, precipitates from etching or washing could be interpreted as additional phases.

*Conventional chemical etching* is the oldest and most commonly applied technique to produce microstructural contrast. In this technique, the etchant reacts with the specimen surface without the use of an external electrical current supply. Etching proceeds by selective dissolution according to the electrochemical characteristics of the component areas *(dissolution etching).*

A contrast can also be developed by material dissolution of the surface and the simultaneous formation of a surface layer; this is for example the case in *precipitation etching.* The material at the surface is dissolved and reacts there with the etchant to form insoluble compounds that deposit onto the sample surface as interference layers or as colored surface coatings. This type of etching is gaining increasing popularity in modern metallography and is also known as *color immersion etching.* Excellent results were obtained with this technique for the examination of grain orientation and inhomogeneities such as coring.

*Heat tinting* is a technique in which an oxide layer is formed by heating the sample in air; since the oxide film forms at different rates on the microstructural phases, the differences in the layer thickness will produce different colors.

Contrary to conventional chemical etching, an electrical potential by means of an external circuit is applied to the sample for *electrolytic or anodic etching.* The same setup used for electrolytic polishing as shown in Fig. 1.10 can be used for electrolytic etching. Both electrodes, the anodic sample and the cathodic counterelectrode, are submersed in a container with the electrolyte (etching solution).

During *anodic etching,* positively charged metal ions are discharged by the specimen surface, leaving behind an equivalent number of electrons in the material that is being removed at the same time. The result is an etching process (as schematically shown in the A–B portion of the current density curve in Fig. 1.11), with actual material removal and without the formation of any layer. This material removal can also be accompanied by forming simultaneously surface layers (portion B–C of the curve in Fig. 1.11). This occurs when the metal ions leaving the specimen surface react with the ions of the electrolyte and form insoluble agents, which are deposited at different thicknesses onto the specimen surface. The thickness of the surface layer varies with the orientation and the composition of the microstructural constituent and appears in their interference colors. This modification of electrolytic etching is also known as *anodizing;* comparable nonelectrolytic techniques are precipitation etching or heat tinting.

*Potentiostatic etching* is an advanced form of electrolytic etching that produces the ultimate etching contrast through highly controlled conditions. The potential of the specimen, which would usually change with the changes of electrolyte concentration, is maintained at a fixed level through the use of a potentiometer and suitable reference standards. Clearly pronounced contrast can be obtained with this method where this is otherwise not possible. In some cases, the cell current can be maintained with a coulombmeter to determine the extent of etching (*reproducible etching,* Fig. 1.16a).

On completion of the etching process, the specimen should be rinsed in clean water to remove the chemicals and stop continuation of any reactions. Sometimes, for example, by etching to reveal segregations using the Oberhoffer method, it is advisable to rinse first with alcohol; otherwise copper could precipitate on the specimen surface because of the change in the degree of dissociation. After the specimens are rinsed in water, they should be rinsed again in alcohol and dried in a stream of warm air (see section 1.6 in this chapter). The use of alcohol improves the drying action and prevents the formation of water stains. If etching produces water-soluble layers, water must be avoided for the rinsing step. Mounted specimens must be cleaned thoroughly to avoid the destructive effects of etchants and solvents seeping from pores, cracks, or clamp interfaces. It may be helpful to use an ultrasonic cleaner to avoid these problems. If specimens are highly porous, or if highly concentrated acids are used for etching (e.g., deep etching), it is advisable to neutralize the chemicals before rinsing and drying the specimen.

Desiccators are best suited for the storage of metallographic samples. Samples can also be protected from atmospheric corrosion with a transparent lacquer. Transparent adhesive tape preserves even very sensitive sample surfaces for many months.

### 1.7.3 Physical Etching

Physical etching methods can also be used to develop microstructural contrast. They leave the surface clean and free from chemical residues; they also offer advantages where electrochemical etching is difficult, for example, when the difference in electrochemical potential among individual constituents is very high or when preparing porous or ceramic materials.

Microstructural contrast produced by *interference layers* is often considered to be an optical method. However, since this technique does not involve a microscopic adjustment, but rather a treatment of the polished surface, it is considered a physical technique.

A transparent layer (film) is deposited by physical methods onto the sample surface. Incident light is partially absorbed by the layer and repeatedly reflected at the layer/specimen interface before it exits. Reflectivity differences (ΔR) of different phases are thus enhanced or lowered. The reflected rays interfere with one another and produce a contrast that appears in color under the microscope (Fig. 1.16b).

The coating techniques for interference layer contrast are vapor deposition and reactive sputtering. Using the *vapor deposition technique,* the coating material is vapor deposited or sublimated as a uniform layer onto the sample surface in an evaporation chamber at a vacuum of $10^{-5}$ mbar. Coating materials for highly reflective surfaces, for example, metals, are typically ZnS, ZnSe, and ZnTe, while cryolite ($Na_3AlF_6$) is used for poorly reflecting surfaces, for example, ceramic materials.

*Reactive sputtering* produces a gas discharge between the sample (anode) and the target (cathode) in an oxygen atmosphere of $5 \times 10^{-5}$ mbar. The positively

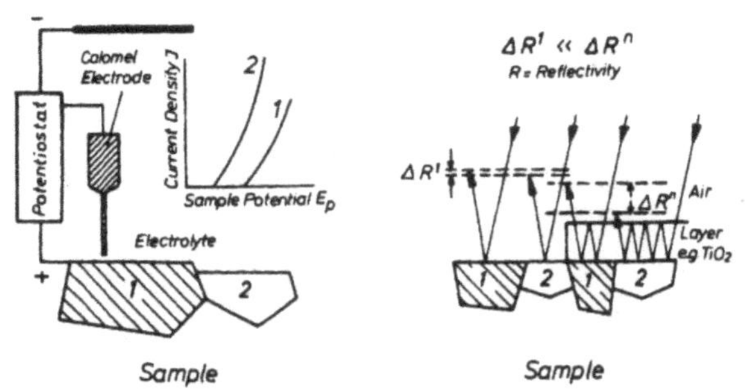


**(a) Electrolytic potentiostatic etching** **(b) Evaporation of interferface layers**

**Fig. 1.16** Methods for reproducible metallographic etching

charged gas ions are accelerated toward the cathode, where they evaporate the cathode material. This in turn reacts with the oxygen and forms a metallic oxide layer on the sample surface. Typical cathode materials are Pt, Fe, Pb, In, or Cu. Specific equipment is available where the sputter coating can be applied with great precision and is controlled microscopically.

If sputtering in an inert gas atmosphere instead of oxygen, the technique is known as *cathodic vaporizing* or *sputtering.* The target material is deposited without reaction as a pure metal layer onto the sample surface. This metal layer has no interference qualities and reflects the light directly. This technique is successfully used for reflective light microscopy and for less reflective surfaces such as ceramic materials. Gold or aluminum are suitable coating materials. A modification of this technique is the *cathodic ion etching,* where the sample is the cathode and ions are expunged from the surface when applying an electric potential. The different rate of removal of the constituents reveals the microstructure.

Other physical etching methods are ion etching and thermal etching.

During *ion etching* the sample surface is bombarded by energy-rich ions in a vacuum. The atoms are expunged from their matrix. This removal produces a contrast as a function of the grain orientation and the phase composition. These techniques are especially of interest for nonconductive materials (see section 2.7.3 in Chapter 2 "Ceramography"). Ion-etching equipment is used with an ion gun and a vacuum of $10^{-4}$ mbar. Figure 1.17 shows the physical etching techniques described above.

*Thermal etching* is mainly used for ceramic materials. This method is based on forming an equilibrium between surface and interface energies that are thermally activated by diffusion processes. This produces curved grain surfaces and grooves at the grain boundaries.

### 1.7.4 Etching Reproducibility

Metallographic etching was considered, to a certain extent, to be empirical and often to be a trial-and-error method. The multitude of etching techniques available, their nonuniform nomenclature, and also the lack of knowledge of the basic processes left this very important part of the metallographic sample preparation somewhat confusing and unclear. The conventional etching, which is most popular for its simplicity and economy, is especially difficult to reproduce because of the electrochemical reactions that are taking place and many other interfering factors of the etchant as well as inhibiting effects on the sample surface (polarization phenomena, overpotential, etc.). To obtain reliable and reproducible etching results, many techniques were developed, and they require specific equipment, which are commercially available. These techniques are *potentiostatic*

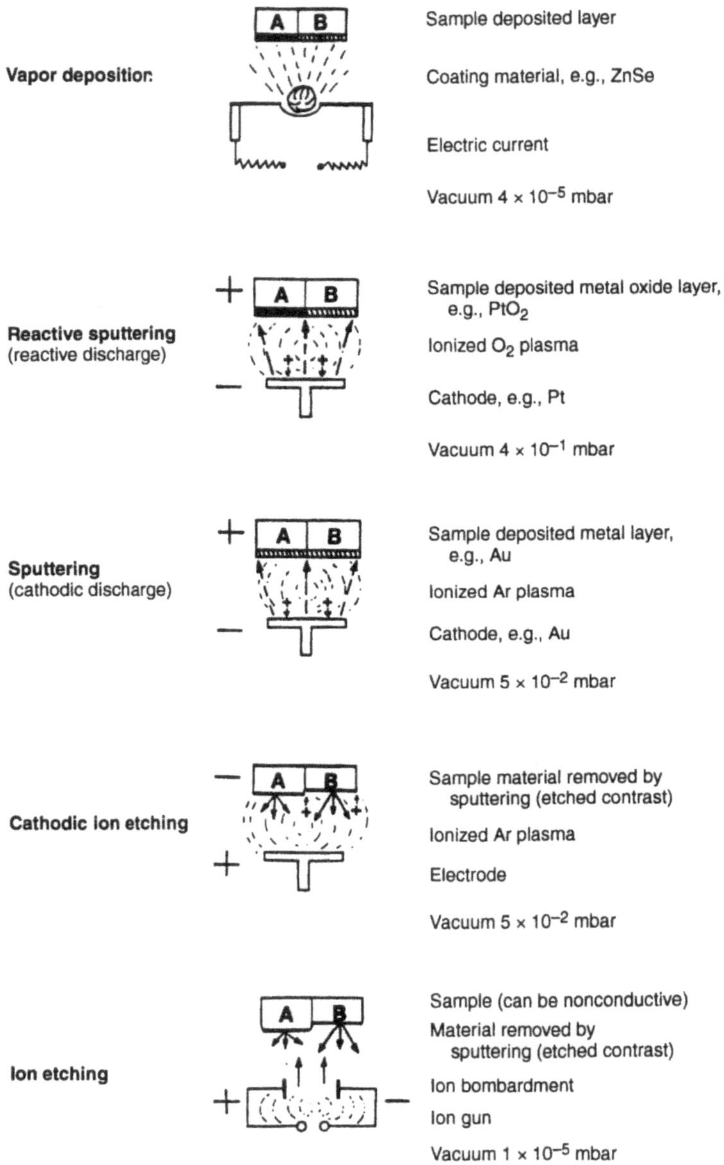


**Fig. 1.17** Techniques for coating of interference layers and for ion etching

*etching* (see section 1.7.2), *ion etching,* and the *deposition of interference layers* (see section 1.7.3 and Fig. 1.16b), which have been described in the previous sections.

Reproducibility in etching is particularly important for *quantitative image analysis.* The numerical evaluation of the microstructure with fully automatic image analysis requires reproducible and uniform contrast of the microstructure, because the equipment is unable to master contrast irregularities and can accumulate erroneous data.

### 1.7.5 Etching Terminology

A classification of the most commonly used terms is possible on the basis of distinct features. A distinction has already been made between optical, electrochemical, and physical etching on the basis of the kinetic phenomena occurring at the sample surface. Other descriptions define either the etching technique, for example, immersion etching, print etching and so forth, or the etching effect, for example, dissolution etching, or the result, for example, grain-contrast etching. Frequently etchants are known by their main ingredient, for example, nital, aqua regia, or they are named after the person who developed the etchant, for example, Adler, Oberhoffer, and so forth. The multitude of names can be quite confusing, especially where several terms are used for the same etching procedure.

A list of etchants with their terminology and explanation is compiled in the following section. An alphabetical list of etchants by their name is listed in section 4.1 "Designation of Etchants," in the Appendix.

### 1.7.6 Definitions of Etching Terms

1. **anodic etching.** Reveals the microstructure by selective anodic dissolution of the polished surface using a DC current. Variation with layer formation: anodizing.
2. **attack polishing.** Simultaneous etching while mechanical polishing.
3. **cathodic etching.** See *ion etching.*
4. **cold etching.** Reveals the microstructure at room temperature and below.
5. **controlled etching.** Electrolytic etching with selection of suitable etchant and voltage, resulting in a balance between current and dissolved metal ions.
6. **crystal figure etching.** Discontinuity in etching depending on crystal orientation. Distinctive sectional figures form at polished surface. Closely related to *dislocation etching.*

7. **deep etching.** Macroetching, especially for steels, to determine the overall character of the material (presence of imperfections such as seam defects, rolling defects, forging bursts, remnant shrinkage voids, cracks, and coring).
8. **dislocation etching.** Reveals exit points of dislocations on the sample surface. Etching of dislocations is caused by their strain field ranging over a distance of several atoms. Crystal figures (etch pits) are formed at the exiting points. For example, etch pits for cubic materials are cube faced.
9. **dissolution etching.** Reveals the microstructure by surface removal.
10. **double etching.** Two etchants are used sequentially, the second one will accentuate a particular microstructural feature.
11. **drop etching.** Placing a drop of an etchant on a selected area of the sample surface to develop the alloying microconstituents (drip reaction).
12. **dry etching.** Develops the microstructure by gaseous exposure.
13. **electrochemical (chemical etching).** General term for revealing the microstructure by redox reactions.
14. **electrolytic etching.** See *anodic etching.*
15. **etch rinsing.** Pouring the etchant over a tilted sample surface until the structure is revealed. Used for etching with severe gas evolution.
16. **eutectic cell etching.** Reveals eutectic grains (cells).
17. **grain-boundary etching.** Reveals the intersections of individual grains. Grain boundaries have a higher dissolution potential than the individual grains because of their high density of structural defects. Accumulation of impurities in grain boundaries increases this effect.
18. **grain-contrast etching.** Etching the surface of the grains according to their crystal orientation. They become distinct by the different reflectivity caused by reaction layers or surface roughness.
19. **heat tinting.** Formation of interference colors in air or other gases, usually at elevated temperature.
20. **hot etching.** Development and stabilization of the microstructure at elevated temperature in etching solutions or gases.
21. **identification (selective) etching.** Etching for the identification of particular microconstituents without attacking any others.
22. **immersion etching.** The sample is immersed in the etchant with the polished surface up and is agitated. This is the most common etching method.
23. **immersion etching (cyclic).** Alternate immersion into two etchants: 1, the actual etchant; 2, solution to dissolve the layer formed during the etching process of 1.
24. **ion etching.** Surface removal by bombardment with accelerated ions in a vacuum (1 to 10 kV).
25. **long-term etching.** Etching times of a few minutes to several hours.

26. **macroetching.** Reveals the macrostructure for the examination with the unaided eye or at a magnification of 50× or less.
27. **microetching.** Reveals the microstructure for microscopic observation at a magnification of 50× or higher.
28. **multiple etching.** A sample is etched sequentially with specific etchants to reveal certain constituents.
29. **network etching.** Formation of networks (subgrain boundaries), especially in mild steels after etching in nitric acid.
30. **optical etching.** Develops the microstructure by using special illumination techniques (darkfield, phase contrast, interference contrast, polarized light).
31. **physical etching.** Develops the microstructure through removal of surface atoms or lowering the grain surface potential.
32. **plasma etching.** High-frequency electromagnetic vibrations produce radicals in a gas mixture that react with the sample surface and cause its removal.
33. **potentiostatic etching.** Anodic development of the microstructure at a constant potential enables a defined etching of singular phases.
34. **precipitation etching.** Develops the microstructure by the formation of reaction products at the sample surface.
35. **primary etching.** Develops the cast microstructures including coring.
36. **print etching (printing).** A carrier material is soaked with an etching solution and is pressed onto the sample surface. The etchant reacts with one of the microstructural constituents forming substances that affect the carrier material. The result is a direct imprint as a life-size image. It is used for the identification of specific elements, for example, sulfur.
37. **secondary etching.** Develops the microstructures that differ from primary structures through transformation and heat treatment in the solid state.
38. **segregation (coring) etching.** Develops segregation (coring) mainly in macrostructures and microstructures of castings.
39. **short-term etching.** Etching time of seconds to a few minutes.
40. **shrink etching.** Produces a precipitate coating on the grain surfaces and shrinks upon drying, generating cracks. Crack orientation depends on the underlying crystal structure.
41. **staining.** Precipitation etching that causes contrast by distinctive staining of microconstituents; different interference colors originate from surface layers of varying thickness. Identifies inhomogeneities.
42. **strain etching.** Reveals regions of high deformation within undeformed areas. Strained areas show increased segregations of precipitates.
43. **swabbing.** Wiping the sample surface with cotton saturated with the etchant; this will simultaneously remove undesired reaction products.

44. **thermal etching.** Annealing the specimen in a vacuum or inert atmosphere. This is a preferred technique for high-temperature microscopy and for ceramics.
45. **wet etching.** The sample surface has been wetted before immersion into the etching solution. This is important when using color etchants.
46. **wipe etching.** See *swabbing*.

## 1.8 Field Metallography/Nondestructive Metallography

Failure analysis must often be carried out on site and on parts that are difficult to transport, such as bridges or other permanent structures, and cannot be sampled in the laboratory. A complete line of portable metallographic equipment is available for grinding, polishing, and etching, and even microscopic examination can be performed on site. An alternative technique is to obtain replicas from the metallographically prepared area on site; these replicas can be examined later in the laboratory and under more favorable conditions.

For rough grinding, grinding and polishing high-powered hand drills are available with attachments for small disks that can be fitted with *grinding paper* of 1200 grit. Other portable equipment has a flexible shaft to drive grinding and polishing disks; larger surface areas, such as crack regions and corrosion layers can be prepared rather quickly this way with conventional methods, for example, grinding with SiC papers and polishing with diamond pastes or alumina. Where the material is known and the microstructure needs to be examined at several different areas (e.g., bulk material, weld joints, etc.), electrolytic polishing is a good alternative. Portable equipment with a power supply, an electrolyte, and polishing container is very useful, especially when the polishing conditions are known.

*Etching* is usually carried out with a cotton swab saturated with the etchant. It is important to work carefully and to abide by all safety precautions. Face guards and gloves should be worn, the surrounding area has to be protected, and a spill of chemicals should be avoided (chemicals should only be transported in containers, and the bottles should always be well closed, etc.). When electrolytically etching, a well-established procedure for the particular material should be used.

For initial microstructural observation, small portable microscopes are available. However, if the areas to be examined are not easily accessible and a photographic documentation of the microstructure is required, a replica of the microstructure can be made and analyzed later in the laboratory.

Three different *replica techniques* are used: One technique is to place a softened foil (generally of an ethyl acetate basis and less than 0.1 mm thick) onto the prepared sample surface, allow it to harden for several minutes, and remove it with tweezers or a double-sided adhesive tape. Another technique uses a two-component, cold-setting polymer that is applied onto the sample surface and removed from the material when it has hardened. The third technique is to place drops of lacquer onto the surface; this may be reinforced by a room-temperature curing polymer.

**Table 1.9 Replica techniques used in field metallography**

| Name (supplier) | Material | Application technique | Optical or electron microscopy |
|---|---|---|---|
| **Bioden** Buehler Dr. Hart | 0.05 mm foil Sheets of 100 × 100 mm Solvent: acetone | Place softened foil flat onto the surface, let dry for a few minutes, remove with tweezers or a double-sided adhesive tape and mount onto an objective slide. Application requires experience. | Versatile application, not suitable for gaps and cavities Excellent replication of pores Excellent replication of grain boundaries and carbides |
| **Triafol BN** Bayer Balzer | 0.1–0.15 mm foil Strip of 50 × 50 mm Solvent: Acetone | Place softened foil flat onto the surface, let dry for a few minutes, remove with tweezers or a double-sided adhesive tape and mount onto an objective slide. Easy to apply | Versatile application, not suited for replication of very rough samples Good replication of cracks and pores Excellent to good replication of grain boundaries |
| **Transcopy** Struers | 0.1–0.15 mm foil, reinforced on one side with aluminum. 10 × 30 mm Solvent: transcopy fluid | Soften the foil surface slightly (not too wet) and press onto the surface with a thumb. Do not move; let harden under pressure for approximately 2 min, remove and mount onto an objective slide. Application takes some skill. | Flat replication of cracks Good to satisfactory replication of the microstructure |
| **Technovit 3040 or 5071** Kulzer | Two components, cold polymerizate. Powder and liquid | Mix two parts of powder with one part of liquid, make a plasticine barrier and pour Technovit. Spray with a cooling spray after hardening and remove with a slight tap. | |
| **Pioloform F50** Wacker Chemie | Dissolve 2% Pioloform in chloroform (lacquer) | Wet fracture surface with the Pioloform (dissolved in chloroform), let dry for approximately 2 min. Apply Technovit as described above and remove carefully by light tapping. | Fracture surface replication. Resolution to below 1 μm |
| **Vernis Replic** Presi Dujardin | Lacquer | Good for round samples, crack propagation, microstructural features, pores Remove with warm water. | Excellent resolution up to 1000× |
| **Xantopren** Bayer | Silicon rubber with activator | Low viscosity, suited for drill holes, cavities, and cracks | Good replication of topography, grain boundaries, and carbides. Pores are not replicated. |

The replica retains details of the surface relief and represents the negative of the microstructure. To improve its contrast, the foil can be sputtered with aluminum or gold. The replica can be examined up to a magnification of 500× times with the optical microscope and up to 5000× with a scanning electron microscope. Replicas can be obtained from areas that are difficult to access, from fracture surfaces, and so forth. The different types of foils and lacquers, their application, and handling are listed in Table 1.9.

Replicating is also a particularly attractive technique for the metallography of radioactive materials. The replicating foil absorbs only a negligible amount of radioactive materials and can therefore be handled from the moment of removal from the sample without the usual necessary elaborate precautions. Another valuable application for replicas is the analysis of fracture surfaces. Replica foils from round surfaces, for example, wires, rods, and so forth, can be flattened for microscopic analysis.

## 1.9 Preparation and Etching Recipes for Metallic Materials

The conditions listed in this section for the preparation and composition of etchants have been well established. They have been revised for this edition and have been updated and tested in our laboratory and in others; however, in most cases they can be easily modified to each particular application for optimum results. As described in the previous sections in this chapter, the sample preparation depends on many different parameters that may have an effect, even when using the most reliable recipes. This can often be improved by trials and variations of the conditions. It is especially applicable to those formulas with the reference "(Concentration can be varied)." Neither do the compositions of the materials exclude that a preparation can still be successful when even larger deviations are made. Incomplete information of the preparation conditions and comments are due to missing information from the literature.

The following sections are subdivided into different material groups I, II, III, and so forth of metals and alloys; materials that require similar preparation techniques are grouped together. Alloy compositions are listed in weight percent.

The recipes listed apply generally for a manual sample preparation. They can naturally be modified for automatic or semiautomatic preparation; in this case the preparation time will be reduced because of the more precise conditions, and the reproducibility will also improve.

Details of the consumables are listed uniformly, for example, SiC grits always refer to the FEPA Standard (Table 1.6); the abrasive grain sizes of alumina are listed in microns. Diamond grits are also listed in microns and can be used as paste, spray, or suspension. Descriptions of polishing disks and cloths are restricted to the type of material and the manufactured product (Table 1.8).

Lubricants are only listed where it is important whether an alcohol- or oil-based lubricant or no water should be used.

Most etching solutions mentioned in the instructions consist of dissolution media containing reactive compounds, for example, distilled water and acid (tap water is rarely suited as a dissolution medium!). The etching reaction can be influenced by substances that increase or decrease the degree of dissolution. The individual components of an etchant have a different effect on different samples as well as different conditions of the samples (for example, cold worked or undeformed).

A large number of etchants are named after their inventors (e.g., Adler), or have traditionally a typical name (e.g., nital etchant). They are listed in section 4.1 "Designation of Etchants," in the Appendix.

Numerous materials and chemicals are dangerous and injurious to health. They are indicated by three exclamation marks (!!!). They are described in full detail in section 4.4, "Handling Tips for Hazardous Materials" in the Appendix.

### *1.9.1 Silver*

Ag

**Preparation**

I. Fine silver: 99.99 to 99.97.

Fine-grain silver: 99.85 with 0.15 Ni; used in chemistry and the electrotechnical industries.

Pure silver is very ductile and difficult to prepare. It smears and deforms easily during grinding and polishing. Abrasive grains tend to become embedded into the surface and are extremely difficult to remove. Grinding and polishing must therefore be carried out with great care. Etching can produce pits on polished surfaces where the deformation has not been entirely removed. Silver forms water-soluble halides, hydroxides, and carbonates; care should be taken when using tap water.

*Grinding:*

a. Wet grinding on fresh SiC papers down to 1200 grit, then continue on microcut paper. Do not skip grit sizes and use little pressure and distilled water.
b. When cutting by ultramilling use a hardmetal or diamond cutter, polishing will not be necessary.

*Polishing:*

Mechanical polishing does not produce satisfactory results.

a. Electrolytic polishing with Ag EW2.
b. Chemical polishing with Ag CP1.

II. Silver-base semifinished products for coins, cutlery, and jewelry, e.g.:

Sterling silver: 92.5Ag-7.5Cu.

835/silver: 83.5Ag-16.5Cu.

800/silver: 80Ag-20Cu.

Coinage silver: 90 to 62.5Ag with Cu.

*Grinding:*

Wet grinding on SiC papers down to 1200 grit.

*Polishing:*

a. On a hard silk cloth with 1 and 0.25 μm diamond particle size. Automatic polishing is recommended because of long polishing times. Manual final polish on a short-nap velvet cloth with a magnesia suspension.
b. Electrolytic lapping for Ag-Au with Ag EW2, for Ag-Pd with Ag EW1.

III. Silver solders: Ag-28Cu, Ag with 33Cd-Cu-Zn.

Brazing solders: 5 to 85 Ag with Cu-Zn.

Contact materials: Ag-Cd, Ag-Ni, Ag-Cd-Ni, Ag-Cu, Ag-Cu-Ni, Ag-W, Ag-CdO, Ag-$SnO_2$, Ag-MgO, Ag graphite.

Ceramic solders: Ag-Ti.

The multiphase solders can be prepared as described under section II, though with greatly reduced times. Sensitive contact solders should be mounted in room-temperature curing material; plated samples should be plated with a metal of a different color for edge reinforcement.

*Grinding:*

Wet grinding on SiC papers down to 1200 grit with light pressure.

*Polishing:*

a. With automatic polishing equipment (if possible), using a hard silk cloth with 1 μm diamond particle size for 30 min, then 0.25 μm diamond particle size for 45 min. Manual final polishing on a short-nap velvet cloth with a magnesia suspension.

b. Electrolytic lapping with Ag EW1.

IV. Precious metals coatings: see section 1.9.3 (Au).

## Electrochemical Polishing Recipes

*Electrolytic Polishing*

| | | | |
|---|---|---|---|
| Ag EP1 | 1000 mL | distilled water | 10 min |
| | 38 g | potassium carbonate | 2.5–4 V DC |
| | 37 g | potassium cyanide | Ag cathode |
| | 35 g | silver cyanide | !!! See section 4.4 in the Appendix. |
| Ag EP2 | 200 mL | distilled water | 4–6 min, 40 °C |
| | 380 mL | ethanol 96% | 25–30 V DC |
| | 400 mL | phosphoric acid 85% | Al cathode |
| Ag EP3 | 1000 mL | distilled water | ~1 min |
| | 100 g | sodium cyanide | 2.5 V DC |
| | 100 g | potassium ferrocyanide | Graphite cathode |
| | | | !!! See section 4.4 in the Appendix. |

*Electrolytic Lapping*

| | | | |
|---|---|---|---|
| Ag EW1 | 1000 mL | distilled water | 3–5 min |
| | 12 g | potassium thiosulfate | 6–8 $mA/m^2$ |
| | 2 g | thiourea | |
| | 1 g | copper (II) nitrate | |
| Ag EW2 | 195 mL | ethylene glycol | 3 min |
| | 25 mL | hydrochloric acid 32% | 25 V |
| | 3 mL | nitric acid 65% | |
| | 30–50 mL | alumina suspension 0.05 μm | |

*Chemical Polishing*

| | | | |
|---|---|---|---|
| Ag CP1 | 800 mL | distilled water | !!! See section 4.4 in the Appendix. |
| | 100 mL | saturated aqueous solution of chromium (VI) oxide | |
| | 45 mL | hydrochloric acid 32% | |

## Macroetching

| No. | Material | Etchant | Conditions |
|---|---|---|---|
| Ag M1 | Pure Ag<br>Grain-contrast etching | 70 mL ethanol 96%<br>30 mL nitric acid 65% | 3–5 min<br>Heat up slightly.<br>!!! See section 4.4 in the Appendix. |
| Ag M2 | Ag alloy with low Cu content<br>Grain-contrast etching | 90 mL ethanol 96%<br>10 mL nitric acid 65% | 1–3 min<br>!!! See section 4.4 in the Appendix. |

## Microetching

| No. | Material | Etchant | Conditions |
|---|---|---|---|
| Ag m1 | Pure Ag | 100 mL distilled water<br>5–10 g potassium cyanide | Several min<br>!!! See section 4.4 in the Appendix. |
| Ag m2 | Pure Ag | 100 mL sulfuric acid 95–97%<br>1–5 g chromium (VI) oxide | Few s<br>Immersion or swab etching<br>!!! See section 4.4 in the Appendix. |
| Ag m3 | Pure Ag and low-alloy Ag | (a) 100 mL distilled water<br>10 mL ammonium persulfate<br>(b) 100 mL distilled water<br>10 g potassium cyanide | 30 s to 2 min<br>(a) and (b): mix just before etching.<br>!!! See section 4.4 in the Appendix. |
| Ag m4 | Pure Ag. Ag-Ni and Ag-Pd alloys | 50 mL ammonia solution 32%<br>50 mL hydrogen peroxide 30% | Few s<br>Mix just before etching.<br>!!! See section 4.4 in the Appendix. |
| Ag m5 | Ag alloys. Silver solders (Ag-Cu-Cd-Zn) | 100 mL potassium chloride solution saturated in cold water<br>10 mL nitric acid 65%<br>2 mL sodium chloride solution saturated | Few s<br>Heat up slightly before etching.<br>!!! See section 4.4 in the Appendix. |
| Ag m6 | Ag-Cd, Ag-Cu. Silver solders (Ag-Cu-Cd-Zn) | 25 mL distilled water<br>25 mL ammonia solution 32%<br>50 mL hydrogen peroxide 30% | Few s<br>Mix before etching!<br>Add hydrogen peroxide at the end.<br>!!! See section 4.4 in the Appendix. |
| Ag m7 | Silver solders (Ag-Cu-Cd-Zn) | 100 mL distilled water<br>2 g iron (III) chloride | 5–30 s<br>Immersion etching or swabbing. |
| Ag m8 | Ag-Cu alloys<br>Grain-boundary etching | 1000 mL distilled water<br>1–2 mL sulfuric acid 95–97%<br>1–2 g chromium (VI) oxide | 1 min<br>!!! See section 4.4 in the Appendix. |
| Ag m9 | Ag-Ni, Ag-Mg-Ni, Ag-Pb-Sn, and Ag-Sn alloys | 100 mL distilled water<br>2–11 mL sulfuric acid 95–97%<br>2 g chromium (VI) oxide<br>(Concentration variable) | Few s<br>!!! See section 4.4 in the Appendix. |
| Ag m10 | Ag-Mo- and Ag-W alloys. Ag-WC | 10 mL aqueous solution of sodium hydroxide 10%<br>10 mL aqueous solution of potassium ferrocyanide 30% | 5–15 s<br>If the attack is too violent, use solution at half strength. |

(continued)

## Microetching (continued)

| No. | Material | Etchant | Conditions |
|---|---|---|---|
| Ag m11 | Ag-Sn-Zn-Cu alloys<br>Dental amalgam:<br>Beta phase is attacked. | 100 mL distilled water<br>3 mL sulfuric acid 95–97%<br>2 g potassium dichromate<br>1 g sodium fluoride<br>(Crowell's reagent) | Seconds to minutes<br>!!! See section 4.4 in the Appendix. |
| Ag m12 | Ag-Th alloys | 100 mL distilled water<br>8 mL sulfuric acid 95–97%<br>2 g potassium dichromate | Seconds to minutes |

## Electrolytic Etching

| No. | Material | Etchant | Conditions |
|---|---|---|---|
| Ag Em1 | Pure Ag | 1000 mL distilled water<br>38 g potassium hydrogen carbonate<br>35 g silver cyanide<br>30 g potassium cyanide | 90 s<br>0.5 V DC<br>Ag cathode<br>!!! See section 4.4 in the Appendix. |
| Ag Em2 | Ag alloys | 1000 mL distilled water<br>100 g citric acid | 15 s to 3 min<br>6 V DC<br>Ag cathode. If necessary, add 2–3 drops of hydrochloric acid. |
| Ag Em3 | Pure Ag. Ag alloys. Ag-Cu alloys | 800 mL distilled water<br>40 g sodium thiosulfate<br>20 g potassium disulfate | 10–20 s<br>Sample must be electrolytically polished.<br>DC. Carbon cathode |

## *1.9.2 Aluminum*

Al

### Preparation

I. High-purity aluminum: 99.98. Pure aluminum: 99.5.

Used for extruded profiles, sheet, foils, and household utensils.

High-purity, pure, and single-phase Al alloys are soft. Sectioning and grinding can cause extensive deformation damage. Loose grit particles embed easily into the sample surface. Revealing the microstructure may be difficult due to the formation of an oxide layer.

*Grinding:*

On SiC papers down to 1200 grit using a lubricating emulsion, petroleum, or soap water (0.5 to 1%). If possible, complete grinding on microcut paper or 2400 grit SiC paper with water as a lubricant; long grinding is necessary, decrease pressure from one step to the next.

*Polishing:*

a. Rough polishing with 5 μm alumina on a felt cloth and at high wheel speed. Final polish on a silk velvet cloth treated with soap, using a thick slurry of MgO or 0.05 μm alumina with an addition of 1% aqueous solution of ammonium tartrate. Caution: the MgO suspension forms carbonates and should only be used for a few minutes. Copper and bronze polishing wheels should not be used because of a possible electrochemical reaction during polishing.

b. Rough polishing with 6 μm diamond particle size on a hard silk cloth and with 1 μm diamond particle size on a short-nap synthetic cloth, using an oil-based lubricant. Final polishing with $SiO_2$ suspension. The suspension should be used as thick and concentrated as possible at the beginning of the final polishing, and then increasingly diluted with water toward the end of this polishing step. The wheel speed should be at 500 rpm with a minimum sample pressure.

c. Diamond polishing down to 3 μm particle size, then shock polishing with Al EP1 for 3 to 5 s. This technique is recommended for single-phase Al-Mg alloys.

d. 6 μm diamond rough polish, followed by an electrolytic polish with Al EP1, Al EP2, Al EP3, or Al EP4.

e. 6 μm diamond polish, followed by a chemical polish with Al CP1.

II. Foundry alloys:

Al-Cu, Al-Si, Al-Si-Cu, Al-Mg, Al-Zn, Al-Sn with additions of Si, Fe, Cu, Mn, Mg, Cr, Ni, Zn, Sn, Ti, and others, as well as Al-Li with Cu, Mg, Zr, Fe, and Si. These alloys are used in machines and engines parts (pistons) and for construction parts.

Foundry alloys are not difficult to prepare, but etching for the identification of the different intermetallic phases is difficult and time consuming since the sample has to be repolished every time for individual phase identification. An easier method for phase identification can be coating the surface with interference layers.

III. Wrought alloys:

Al-Cu, Al-Mn, Al-Si, Al-Mg, Al-Mg-Si, and Al-Zn with alloying elements of Si, Fe, Cu, Mn, Mg, Cr, Ni, Zn, Ga, Ti, and others are used for extrusions, sheet material, and aluminum siding for buildings.

Wrought alloys are slightly harder than foundry alloys, but can be prepared with the same method.

*Grinding:*

Wet grinding on SiC papers down to 1200 grit, using little pressure; then continue with abrasive cloth.

*Polishing:*

a. With 6 μm diamond particle size on a hard synthetic cloth, 3 μm diamond particle size on cotton cloth, and 1 μm diamond particle size on a short-nap synthetic cloth using an oil-based lubricant. If a color etchant is to be used, final polish with $SiO_2$ suspension on a long-nap velvet cloth. Cu-containing alloys should not be polished together with non-Cu-containing alloys. Ag-containing alloys should be final polished with 0.25 μm diamond particle size on a short-nap velvet cloth using an oil-based lubricant. Mg-containing phases tend to tarnish and corrode; they are best polished without water! Soft phases such as lead are easily removed during polishing; a cotton cloth is suitable and a short polishing time of 1 to 3 min is recommended, **do not use any water.**
b. Use ultramilling only for harder alloys.
c. Electrolytic polishing with Al EP1, Al EP2, Al EP3, or Al EP4 does not give good results for multiphase alloys, and mechanical polishing is preferable. The sample surface should be rough polished with 3 μm diamond particle size. Short polishing times are recommended to avoid relief formation.
d. Rough polishing with 3 μm diamond particle size and subsequent shock polishing with Al EP1 gives good results for Al-Mg and Al-Mg-Si alloys.
e. Al-Mg-Cu alloys should be final polished with very fine alumina, treated with soap and HF water (5 drops of HF per 500 mL distilled water) on a long-nap velvet cloth (use gloves!) and at high wheel speeds.

IV. Aluminum metal-matrix composites:

Al-Si-Cu with 10 to 20 vol% SiC or $Al_2O_3$ used in the automotive, aerospace, and the weapons industries.

Hard ceramic particles are distributed in the soft aluminum matrix. A manual preparation is not recommended, and the samples should be ground and polished with automatic polishing equipment.

*Grinding:*

On a 20 μm (or coarser) diamond disk with water as the lubricant. Pressure used for 6 samples = 120 N, rotating wheel speed = 150 rpm.

*Lapping:*

On a lapping disk with 15 μm diamond particle size, 90 N pressure, at 150 rpm for approximately 5 min.

*Polishing:*

On a perforated synthetic cloth with 15 μm diamond particle size, 120 N pressure, at 150 rpm for 2 min; then with 6 μm diamond particle size also on perforated synthetic cloth at 120 N pressure, at 150 rpm for 2 min. This is followed by a 1 μm diamond polish on a hard synthetic cloth with 90 N pressure at 150 rpm; final polish on a short-nap synthetic cloth with $SiO_2$ suspension using 120 N pressure at 150 rpm for 30 to 60 s.

**Electrochemical Polishing Methods**

*Electrolytic Polishing*

| | | |
|---|---|---|
| Al<br>EP1 | 200 mL perchloric acid 60%<br>700 mL ethanol 96%<br>100 mL butyl cellosolve 99%<br>(20–30 °C) | Water cooling<br>Stainless steel cathode<br>!!! See section 4.4 in the Appendix. |
| Al<br>EP2 | 140 mL distilled water<br>800 mL ethanol 96%<br>60 mL perchloric acid 60% | 15–60 s<br>30–80 V DC<br>Stainless steel cathode<br>!!! See section 4.4 in the Appendix. |
| Al<br>EP3 | 250 mL distilled water<br>380 mL ethanol 96%<br>400 mL phosphoric acid 85% | 4–6 min, 40 °C<br>40–60 V DC<br>Al cathode |
| Al<br>EP4 | 220 mL distilled water<br>200 mL glycerol 87%<br>580 mL sulfuric acid 95–97% | 1–20 min<br>1.5–12 V DC<br>Stainless steel cathode<br>!!! See section 4.4 in the Appendix. |

*Chemical Polishing*

| | | |
|---|---|---|
| Al<br>CP1 | 25 mL sulfuric acid 95–97%<br>70 mL phosphoric acid 85%<br>5 mL nitric acid 65% | 30 s to 2 min<br>85 °C |

*Electrolytic Lapping*

| | | |
|---|---|---|
| Al<br>EWP1 | 95 mL distilled water<br>10 mL alumina suspension 0.05 μm<br>5 mL sulfuric acid 95–97%<br>0.5 g sodium fluoride | 2–6 min<br>1–10 mA/6.5 $cm^2$ DC<br>Change polarity every 1.5 s. |

## Macroetching

Grinding on SiC papers down to 800 grit is sufficient.

| No. | Material | Etchant | Conditions |
|---|---|---|---|
| Al M1 | Al 99.99–98 and all alloys<br>Reveals oxides, cracks, etc. | 100 mL distilled water<br>5–20 g sodium hydroxide | 5 min to 1 h<br>At RT or up to 50 °C |
| Al M2 | Cu-, Mn-, Si-, Mg-, and Ti-containing Al alloys<br>Foundry alloys with high Si content | 75 mL hydrochloric acid 32%<br>25 mL nitric acid 65%<br>5 mL hydrofluoric acid 40% | Seconds to minutes<br>Use only fresh solution. If necessary dilute with 25 mL distilled water. Rinse sample with warm water.<br>!!! See section 4.4 in the Appendix. |
| Al M3 | Al alloys<br>Pure Al | 90 mL distilled water<br>15 mL hydrochloric acid 32%<br>10 mL hydrofluoric acid 40%<br>(Flick's etchant) | 5 s to 3 min<br>!!! See section 4.4 in the Appendix. |
| Al M4 | Common Al alloys and pure Al<br>Al-Mn, Al-Mg, Al-Mg-Mn, and Al-Mg-Si alloys<br>Grain size. Rolling texture. Welds<br>Is also used as a microetchant | 20 (50) mL distilled water<br>20 (15) mL hydrochloric acid 32%<br>20 (25) mL nitric acid 65%<br>5 (10) mL hydrofluoric acid 40%<br>(Keller's etchant) | 1–3 min<br>!!! See section 4.4 in the Appendix. |
| Al M5 | Pure Al. Mn-Al, Si-Al, Mg-Al, and Mg-Si-Al alloys | 25 mL distilled water<br>45 mL hydrochloric acid 32%<br>15 mL nitric acid 65%<br>15 mL hydrofluoric acid 40%<br>(Tucker's etchant) | Seconds to minutes<br>Use fresh solution.<br>!!! See section 4.4 in the Appendix. |
| Al M6 | Surface defects<br>Pure Al | 100 mL distilled water<br>5.5 mL sulfuric acid 95–97% | Seconds to minutes |

If the surface becomes smudged after macroetching, remove with an aqueous solution of nitric acid (1:1).

## Microetching

| No. | Material | Etchant | Conditions |
|---|---|---|---|
| Al m1 | For most Al type alloys<br>Al-Be alloys. Grain boundaries<br>Slip lines in pure Al | 100 mL distilled water<br>0.5 mL hydrofluoric acid 40%<br>If necessary, increase hydrofluoric acid to 10 mL. | 10–60 s<br>Etch and polish alternately.<br>!!! See section 4.4 in the Appendix. |
| Al m2 | For most Al types and alloys, except for Al alloys with high Si content<br>Al foils | 95 mL distilled water<br>2.5 mL nitric acid 65%<br>1.5 mL hydrochloric acid 32%<br>1 mL hydrofluoric acid 40%<br>(Dix and Keller's reagent)<br>(Concentration variable) | 10–30 s<br>Use fresh solution.<br>!!! See section 4.4 in the Appendix. |
| Al m3 | Pure Al. Cu-Al, Mg-Al, Mg-Si-Al, and Zn-Al alloys<br>The $Al_2Cu$ phase remains unetched after very short etching times.<br>Shrink etching | (a) 100 mL distilled water<br>1–2 g sodium hydroxide<br>(b) 95 mL distilled water<br>5 mL nitric acid 65% | Etch 5–10 s in (a). If necessary heat to 50 °C. Use only fresh solution!<br>Rinse in (b).<br>Etch again for 15 min in (a) and rinse in water for approximately 10 min. |
| Al m4 | For most Al materials and Al alloys | 100 mL distilled water up to 25 g sodium hydroxide<br>1 g zinc chloride | Few seconds to minutes |
| Al m5 | For most Al types and alloys, especially Cu-containing alloys. AlZnMg secondary precipitates. | 75 mL distilled water<br>25 mL nitric acid 65% | 40 s at 70 °C<br>1–2 h at RT |
| Al m6 | Suited especially for Al-Cu alloys (also a macroetchant) | 92 mL distilled water<br>6 mL nitric acid 65%<br>2 mL hydrofluoric acid 40%<br>(Kroll's reagent) | 15 s<br>!!! See section 4.4 in the Appendix. |
| Al m7 | Identification of intermetallic phases in Al alloys containing high concentrations of Cu, Mn, Mg, Fe, Be, Ti, $Fe_3Al$<br>Al-Fe and Al-Fe-Si phases<br>Secondary precipitates | 80 mL distilled water<br>20 mL sulfuric acid 95–97% | 30 s to 3 min at 70 °C |
| Al m8 | Evidence of susceptibility for intercrystalline corrosion in Al-Mg alloys. Grain boundaries are clearly delineated in alloys susceptible to intercrystalline corrosion.<br>Precipitates of $Al_8Mg_5$ or $Al_3Mg_2$.<br>Lüders bands in Al- Mg-base alloys | 100 mL distilled water<br>9 g phosphoric acid (crystalline) | 30 min at 20 °C<br>Before etching, heat sample to 100 °C and cool slowly. |
| Al m9 | Pure Al. Al-Mg and Al-Mg-Si alloys<br>Etch pits | 25 (50) mL methanol 99.8%<br>25 (30) mL hydrochloric acid 32%<br>25 (20) mL nitric acid 65%<br>1 drop hydrofluoric acid 40% | 10–60 s at 20–25 °C<br>!!! See section 4.4 in the Appendix. |

(continued)

## Microetching (continued)

| No. | Material | Etchant | Conditions |
|---|---|---|---|
| Al m10 | Al-Si and Al-Cu alloys<br>Segregations<br>Grain boundaries | 60 mL distilled water<br>10 g sodium hydroxide<br>5 g potassium ferricyanide | 2 min |
| Al m11 | Pure Al. Al-Cu, Al-Mg, and Al-Mg-Si alloys | 90 mL distilled water<br>10 mL phosphoric acid 85% | (a) 1–3 min at 50 °C<br>(b) Electrolytic: 5–10 s 1–8 V DC<br>Stainless steel cathode |
| Al m12 | Sintered Al materials<br>Grain boundaries<br>Removes smeared layers | 84 mL distilled water<br>15.5 mL nitric acid 65%<br>0.5 mL hydrofluoric acid 40%<br>3.0 g chromium (VI) oxide | 20–60 s<br>Use fresh solution!<br>!!! See section 4.4 in the Appendix. |
| Al m13 | Al-Si-Cu foundry alloys<br>Color etchant | 200 mL distilled water<br>1 g ammonium molybdate<br>6 g ammonium chloride | Etch at 30 s intervals. |
| Al m14 | Reveals coring in foundry alloys<br>Welds<br>Color etchant | 100 mL distilled water, 40 °C<br>4 g potassium permanganate<br>1 g sodium hydroxide | 7–20 s<br>Nonhardenable alloys for 45 s |
| Al m15 | High-purity Al<br>Most Al alloys | (a) Add 5 mL hydrofluoric acid 40% to 95 mL distilled water and bring to boiling point. Add enough crystalline molybdic acid to saturate the solution.<br><br>If necessary use combined with: | Few s for alloys; min for pure Al<br>Sample must be wetted before etching.<br>Etch at RT. |
| | Grain boundaries are preferentially attacked. | (b) Add 10 mL hydrofluoric acid 40% to 90 mL distilled water, bring to boiling point and saturate the solution with molybdic acid. | As for (a)<br>Cu- and Zn-containing alloys will discolor dark.<br>Use (b) first and then (a).<br>!!! See section 4.4 in the Appendix. |

## Electrolytic Etching

| No. | Material | Etchant | Conditions |
|---|---|---|---|
| Al Em1 | Pure Al. Al-Zn, Al-Mn, Al-Mg-Si, Al-Zn-Mg and Al-Mn-Mg alloys<br>Marginal results for Al-Cu-Mg alloys | 200 mL distilled water<br>5 mL fluoroboric acid 35% (Barker's etchant) | 1–2 min<br>20–40 V DC. Do not wipe surface during rinsing.<br>Al, Pb, or stainless steel cathode<br>Examine under polarized light.<br>!!! See section 4.4 in the Appendix. |
| Al Em2 | Pure Al. Al-Cu, Al-Mg, and Al-Mg-Si alloys | 90 mL distilled water<br>10 mL phosphoric acid 85% | 5–10 s<br>1–8 V DC current<br>Stainless steel cathode |

**Physical Etching**

Coating the samples with interference layers may help identify and distinguish many, though not all, of the numerous intermetallic phases. Low absorption ZnTe coatings with a high refractive index are commonly used; they provide good color contrast for the matrix as well as for the low reflecting phases. ZnSe, ZnS, and also reactive sputtering of $Fe/O_2$ are producing weak colors for the matrix, but strong colors for the intermetallic phases. The quality of the final polish affects the reflectivity of the phases. Best results are obtained with an electrolytic polish followed by a final polish with 0.25 μm diamond particle size.

Au

### *1.9.3 Gold*

**Preparation**

I. Pure gold, 99.9 (24 karat).

For jewelry. In the electronic industry, e.g., contact wires.

Ducat gold AuCu 1.4.

Pure gold is soft; to remove very fine scratches, the sample has to be polished electrochemically or with a combined chemical-mechanical polish. The sample should then be mounted in a suitable conductive material.

*Grinding:*

a. Wet grinding on SiC papers down to 2400 grit without pressure.
b. Ultramilling with a diamond knife.

*Polishing:*

a. With 6 μm diamond particle size on a hard synthetic cloth, then with 1 μm diamond particle size on a short-nap velvet cloth.

Final polish: Etch polish on a felt cloth using a diluted alumina suspension (0.05 μm) with Au m4.

b. As for I.a up to 1 μm, then electrolytic lapping with Au EW1.
c. As for I.a up to 1 μm, then electrolytic polish with Au EP1.

II. Au-Ag and Au-Pt alloys for applications in the chemical industry.

Au-Ag, Au-Cu, Au-Ag-Cu, Au-Co, Au-Ni, Au-Cd-Cu alloys for the electronics industry.

Mint gold: Au-10Cu.

Yellow gold: 33.3 Au (8 karat). 58.5 Au (14 karat). 75 Au (18 karat). 83.3 Au (20 karat) with Cu and Ag for jewelry.

White gold: 58.5 Au and 75 Au with Ni, Cu, Zn or Pd, Cu, Zn, Ag, for jewelry.

Solders: 33 Au, 58.5 Au, 70 Au with Ag, Cd, Cu, Zn. 58.5 Au, 75 Au with Ni or Pd, Ag, Cd, Cu, Zn.

Dental alloys: Au with additions of Pt, Pd, Ag, and other metals.

*Grinding:*

a. Wet grinding on SiC papers down to 2400 grit.
b. Ultramilling with a diamond knife for all alloys with a Vickers hardness of less than 150.

*Polishing:*

a. With 6 μm diamond particle size on a hard synthetic cloth, then with 1 μm diamond particle size on a short-nap velvet cloth.

Final polish: Etch polish on a felt cloth with diluted alumina suspension (0.05 μm) and Au m4.

b. As for II.a to 1 μm, then electrolytic lapping with Au EW1.
c. As for II.a to 1 μm, then electrolytic polish with Au EP1.

III. Precious metals as coatings on ordinary nonprecious metals, ceramics, and glasses either plated, welded, or soldered; galvanic or electroless; or vacuum deposited.

Main interests are the coating thickness and the adhesion or bonding of the coating. Generally, the substrate and the coating have different hardnesses. It is therefore recommended to prepare the samples with automatic sample preparation equipment, which provides uniform and even sample pressure. If edge retention is important, the coating should be reinforced with a support layer of a different color (see section 1.3.3).

*Grinding:*

a. With SiC papers down to 1200 grit. If the substrate material is very hard grind flat on a 10 μm diamond disk.
b. Lapping on a medium-hard lapping disk with 3 μm diamond particle size, or with 6 and 3 μm diamond on a perforated pellon cloth using an oil-based lubricant.

*Polishing:*

With 1 μm diamond particle size on a hard silk cloth, using high pressure and short polishing times (1–2 min), then with 0.25 μm on a short-nap cloth using medium pressure and short times, and using an oil-based lubricant.

Final polish: To eliminate very fine scratches of the coating, etch polish with alumina (0.05 μm) and a few drops of the etching solution Au m4.

## Electrochemical Polishing Recipes

*Electrolytic Polishing*

| | | | |
|---|---|---|---|
| Au EP1 | 1000 mL | distilled water | 2–4 min |
| | 80 g | potassium cyanide | 7.5 V DC |
| | 40 g | potassium carbonate | Graphite cathode |
| | 60 g | gold (III) chloride | !!! See section 4.4 in the Appendix. |

*Electrolytic Lapping*

| | | | |
|---|---|---|---|
| Au EW1 | 940 mL | distilled water | 4–5 min |
| | 20 g | potassium thiocyanate | 55 mA/cm$^2$ DC |
| | 20 g | ammonium chloride | |
| | 20 g | ammonium thiosulfate | |

## Macroetching

| No. | Material | Etchant | Conditions |
|---|---|---|---|
| Au M1 | Pure Au | 66 mL hydrochloric acid 32%<br>34 mL nitric acid 65% | Seconds to minutes<br>Etch hot!<br>Use only fresh!<br>!!! See section 4.4 in the Appendix. |

## Microetching

| No. | Material | Etchant | Conditions |
|---|---|---|---|
| Au m1 | Pure Au<br>Au-Pt and Au-Ag alloys with more than 80% Au | 60 mL hydrochloric acid 32%<br>20 mL nitric acid 65%<br>(Aqua regia)<br>(Concentration variable) | Seconds to minutes<br>Use only fresh!<br>Heat solution if necessary.<br>!!! See section 4.4 in the Appendix. |

(continued)

## Microetching (continued)

| No. | Material | Etchant | | Conditions |
|---|---|---|---|---|
| Au m2 | Au alloys with precious metal content of less than 90%<br>Au alloys with a high precious metal content<br>White gold | (a) 100 mL<br>10 g<br>(b) 100 mL<br>10 g | distilled water<br>potassium cyanide<br>distilled water<br>ammonium persulfate | 30 s to 2 min<br>Mix a + b at 1:1 ratio.<br>Use double amount of potassium cyanide and ammonium persulfate.<br>!!! See section 4.4 in the Appendix. |
| Au m3 | Au-Cu-Ag alloys | 100 mL<br>100 mL<br>32 g | distilled water<br>aqueous solution of hydrogen peroxide 3%<br>iron (III) chloride | Seconds to minutes |
| Au m4 | Pure Au and Au-rich alloys | 100 mL<br>1–5 g | hydrochloric acid 32%<br>chromium (VI) oxide | Seconds to minutes<br>Swab etching. Use to remove deformation layers during polishing.<br>!!! See section 4.4 in the Appendix. |

## Electrolytic Etching

| No. | Material | Etchant | | Conditions |
|---|---|---|---|---|
| Au Em1 | Au and Au alloys | 100 mL<br>5 g | distilled water<br>potassium cyanide | 1–2 min. 1–5 V AC<br>Pt or graphite cathode, 0.5–1.5 A/cm$^2$<br>!!! See section 4.4 in the Appendix. |
| Au Em2 | Pure Au | | Hydrochloric acid 32% | 1–2 min<br>5 V AC. Pt cathode |

Be

### *1.9.4 Beryllium*

**Preparation**

Beryllium is toxic when inhaled. All preparatory steps generating toxic dust must be performed in a glovebox and under a fume hood!

I. Single crystals (for neutron and synchroton monochromators). Pure beryllium.

Beryllium tends to cleavage fracture easily; this tendency is directly related to the degree of contamination and grain size. Grinding must be performed with great care and very little pressure.

*Grinding:*

Wet grinding on SiC papers down to 1200 grit. Single crystals should be ground only in one direction using a stationary grinder.

*Polishing:*

Electrolytic with electrolyte Be EP1.

II. Technical Be-type alloys (powder metallurgical products containing oxides), e.g., for x-ray windows, and satellite and rocket parts.

*Grinding:*

Wet grinding on SiC papers down to 1200 grit. Final grinding on 1200 grit SiC paper prepared with paraffin or on abrasive cloth.

*Polishing:*

a. Diamond polishing with 6 μm particle size on a hard synthetic cloth, and then with 3 and 1 μm on cotton cloth. Final polishing with alumina slurry (0.05 μm) and a few drops of ammonia solution, or with a $SiO_2$ suspension.
b. 6 μm diamond polish, then vibratory polishing for 2 to 6 h with 0.05 μm alumina and some addition of $H_2O_2$.
c. Shock polishing with electrolyte Be EP1.

III. Be-containing alloys (alloying elements: Cu, Co, Ni, Fe), e.g., for spring materials and nonsparking tools.

*Grinding:*

Wet grinding on SiC papers down to 1200 grit. Final grinding on 1200 grit SiC paper prepared with paraffin or on abrasive cloth.

*Polishing:*

a. Diamond polishing with 6 μm particle size on a hard synthetic cloth; then with 3 and 1 μm diamond particle size on cotton cloth. Final polish with alumina slurry (0.05 μm) and some drops of ammonia solution or with $SiO_2$ suspension.
b. 6 μm diamond polish, then vibratory polishing for 2 to 6 h with 0.05 μm alumina slurry with some $H_2O_2$.

## Electrochemical Polishing Recipes

*Electrolytic Polishing*

| | | |
|---|---|---|
| Be EP1 | 800 mL ethylene glycol<br>80 mL nitric acid 65%<br>16 mL sulfuric acid 95–97%<br>16 mL hydrochloric acid 32% | 2–3 min<br>RT, 1.8 A/cm²<br>20 V, stainless steel cathode<br>!!! See section 4.4 in the Appendix.<br>!Explosive! |

*Electrolytic Lapping*

| | | |
|---|---|---|
| Be<br>EW1 | 97 mL ethylene glycol<br>1 mL hydrochloric acid 32%<br>2 mL nitric acid 65% | 1–1.5 A/cm²<br>Hard silk cloth |

## Macroetching

Grinding on SiC paper down to 1000 grit is sufficient.

| No. | Material | Etchant | Conditions |
|---|---|---|---|
| Be M1 | Technical Be-type materials<br>Coarse-grained Be | 90 mL distilled water<br>10 mL hydrochloric acid 32%<br>4 g ammonium chloride | Few min<br>Swabbing or immersion etching |
| Be M2 | Technical Be-type materials; especially coarse-grained and low-alloy Be | 90 mL distilled water<br>10 mL hydrochloric acid 32%<br>2 g ammonium chloride<br>2 g picric acid | Few min<br>Swabbing or immersion etching<br>!!! See section 4.4 in the Appendix. |
| Be M3 | Pure or technical grade Be | 80 mL distilled water<br>14 mL sulfuric acid 95–97%<br>6 mL phosphoric acid 85%<br>43 g chromium (VI) oxide | 3 min<br>40 °C<br>!!! See section 4.4 in the Appendix. |
| Be M4 | Pure Be<br>Fine-grained Be<br>Single crystals | 50 mL distilled water<br>40 mL nitric acid 65%<br>10 mL hydrofluoric acid 40% | Seconds to minutes<br>!!! See section 4.4 in the Appendix. |

## Microetching

| No. | Material | Etchant | Conditions |
|---|---|---|---|
| Be m1 | Pure Be<br>Grain boundaries, impurities, single crystals | 50 mL distilled water<br>50 mL nitric acid 65%<br>Some drops hydrofluoric acid 40% | Seconds to minutes<br>!!! See section 4.4 in the Appendix. |
| Be m2 | Pure Be<br>Etch pits | 100 mL distilled water<br>20 g chromium (VI) oxide<br>14 mL phosphoric acid 85%<br>1 mL sulfuric acid 95–97% | 10 s<br>!!! See section 4.4 in the Appendix. |
| Be m3 | Precipitates<br>Grain boundaries | 100 mL distilled water<br>10 g oxalic acid | 2–15 min<br>Boiling<br>!!! See section 4.4 in the Appendix. |
| Be m4 | For most Be material types and Be alloys | 100 mL distilled water<br>5 mL sulfuric acid 95–97% | 1–15 s |
| Be m5 | Be and Be alloys | 25 mL glycerol 87%<br>5 mL hydrofluoric acid 40%<br>5 mL nitric acid 65% | ~15 s<br>If necessary, increase hydrofluoric acid to 15 mL.<br>!!! See section 4.4 in the Appendix. |

(continued)

## Microetching (continued)

| No. | Material | Etchant | Conditions |
|---|---|---|---|
| Be m6 | Be-Co alloys | 15 mL distilled water<br>15 mL glacial acetic acid<br>60 mL hydrochloric acid 32%<br>15 mL nitric acid 65% | Several s<br>!!! See section 4.4 in the Appendix. |
| Be m7 | Be-Cu alloys with high Cu content<br>Be-Ag and Be-Al-Ti alloys | 50 mL distilled water<br>20 mL ammonia solution 25%<br>3 mL hydrogen peroxide 30% | Seconds to minutes<br>Mixture can be varied. Use fresh only! |
| Be m8 | Be-Fe alloys | 100 mL ethanol 96%<br>5 mL hydrochloric acid 32%<br>5 g picric acid | Seconds to minutes<br>!!! See section 4.4 in the Appendix. |
| Be m9 | Be-U alloys!!!<br>Be-Nb, Be-Y, and Be-Zr alloys | 50 mL lactic acid 90%<br>50 mL nitric acid 65%<br>50 mL hydrofluoric acid 40% | Seconds to minutes<br>!!! See section 4.4 in the Appendix. |
| Be m10 | Precipitates | Saturated aqueous copper (II) sulfate solution | 30 s |

## Electrolytic Etching

| No. | Material | Etchant | Conditions |
|---|---|---|---|
| Be Em1 | Be and Be alloys | 294 mL ethylene glycol<br>4 mL hydrochloric acid 32%<br>2 mL nitric acid 65% | 6 min<br>30 °C<br>13–20 V DC<br>Stainless steel cathode |

## Physical Etching

Pure Be and Be solid solutions are anisotropic and can be examined unetched under polarized light. Contrast can be improved by reactive coatings with a Pt cathode and oxygen.

Bi

Sb

### *1.9.5 Bismuth and Antimony*

## Preparation

I. Pure bismuth. Pure antimony. Both metals are primarily used as alloying elements.

Bi and Sb are very brittle, and they should be sectioned with great care; since they fracture very easily, it is recommended that samples be mounted before sectioning.

*Grinding:*

Wet grinding on SiC papers down to 1200 grit with ample water, then with microcut paper.

*Polishing:*

a. With 6 μm diamond particle size followed by a chemical-mechanical polish on a vibratory polisher with 0.05 μm alumina slurry to which a solution of 3% nitric acid in glycerol is added.
b. Electrolytic polishing. For Bi: Bi EP1 and Bi EP2.

II. Bi alloys such as Bi-Sn, Bi-Sb, and Bi-Pb-Sn, e.g., Wood's Metal: 50Bi-25Pb-12.5Sn-12.5Cd.

Quick solders such as 50Bi-10Cd-27Pb-13Sn.

*Grinding:*

Wet grinding on SiC papers down to 1200 grit.

*Polishing:*

With 6 μm diamond particle size and 1 μm on cotton cloth; final polishing with 0.05 μm alumina slurry on a velvet cloth.

III. Sb alloys see as for lead alloys under III. Sb is used as an alloying element in low concentrations (higher levels will cause embrittlement).

*Electrolytic Polishing*

| | | |
|---|---|---|
| Bi EP1 | 75 mL glycerol 87%<br>12.5 mL glacial acetic acid<br>2.5 mL nitric acid 65% | 1–5 min, RT<br>12 V DC<br>Stainless steel cathode<br>Do not store!<br>!!! See section 4.4 in the Appendix. |
| Bi EP2 | 98 mL saturated aqueous potassium iodine solution<br>2 mL hydrochloric acid 32% | 30 s, several times<br>5–7 V DC. 0.2 A/cm²<br>Pb cathode |

## Macroetching

| No. | Material | Etchant | Conditions |
|---|---|---|---|
| Bi M1 | Technically pure Sb<br>Crystal orientation.<br>Sb-Bi alloys | (a) 220 mL distilled water<br>80 mL nitric acid 65%<br>(b) 300 mL distilled water<br>45 g ammonium heptamolybdate | Seconds to minutes<br>Mix even portions of (a) and (b) before using. |
| Bi M2 | Crystal orientation<br>Casting defects in Sb and Bi | 100 mL distilled water<br>25 g citric acid<br>10 g ammonium heptamolybdate | Seconds to minutes |

(continued)

## Macroetching (continued)

| No. | Material | Etchant | Conditions |
|---|---|---|---|
| Bi M3 | Bi-Sn and Sb-Pb alloys | (a) 160 mL distilled water<br>40 mL nitric acid 65%<br>30 mL glacial acetic acid<br>(b) 400 mL distilled water<br>1 mL glacial acetic acid | Etch first with (a) at 40 °C. Repolish until surface is bright. Then etch with solution (b) for 1–2 h. |

## Microetching

| No. | Material | Etchant | Conditions |
|---|---|---|---|
| Bi m1 | For pure and technically pure Sb materials and low-alloy Sb | 70 mL distilled water<br>30 mL hydrochloric acid 32%<br>5 mL hydrogen peroxide 30% | Seconds to minutes |
| Bi m2 | Sb and Sb alloys | 30 mL glacial acetic acid<br>10 mL hydrogen peroxide 30% | Seconds to minutes |
| Bi m3 | Sb and Sb alloys<br>Pb-Sb, Bi-Sn, and Bi-Cd alloys | 100 mL distilled water<br>30 mL hydrochloric acid 32%<br>2 g iron (III) chloride | Seconds to minutes |
| Bi m4 | Sb alloys<br>Grain-contrast etchant | 30 mL distilled water<br>15 mL hydrochloric acid 32%<br>50 mL aqueous solution of sodium thiosulfate 16%<br>3 mL aqueous solution of chromium (VI) oxide 10%<br>(Czochralski reagent) | Add chromium (VI) oxide solution just before using. Chromium (VI) oxide solution content can be increased.<br>!!! See section 4.4 in the Appendix. |
| Bi m5 | Sb, Bi, and their alloys | 50 mL distilled water<br>50 mL hydrochloric acid 32% | 1–10 min |
| Bi m6 | Bi-Sn eutectic, Bi-Cd, and Bi-Pb alloys | 95 mL ethanol 96%<br>5 mL nitric acid 65% | Seconds to minutes |
| Bi m7 | Sb-Pb alloys | 100 (100) mL glycerol 87%<br>25 (9) mL nitric acid 65%<br>25 (9) mL glacial acetic acid<br>(Concentration variable) | Seconds to minutes<br>!!! See section 4.4 in the Appendix. |
| Bi m8 | Bi | 100 mL distilled water<br>5 g silver nitrate | Seconds to minutes |
| Bi m9 | Bi and Sb alloys | 50 mL distilled water<br>50 mL hydrochloric acid 32% | 1–10 min |
| Bi m10 | Bi-In alloys | 97 mL distilled water<br>3 mL hydrochloric acid 32% | Seconds to minutes |

Cd
In
Tl

### *1.9.6 Cadmium, Indium, and Thallium*

#### Preparation

Cadmium and thallium are toxic (!!! See section 4.4 in the Appendix.).

I. Pure cadmium, indium, and thallium.

Pure Cd, In, and Tl are very soft, and sample preparation is very difficult. The three pure metals have a very low recrystallization temperature; pressure and heat generation should therefore be kept at a minimum to avoid deformation and smeared layers or ghost structures.

*Mounting:*

Use only cold-setting mounting.

*Grinding:*

a. Use a special file for light and soft metals.
b. Wet grinding on SiC papers down to 1200 grit, coat papers with paraffin or stearin if necessary.
c. Ultramilling with a hardmetal or diamond knife, this saves polishing; the sample should be well cooled during the milling operation.
d. Machining, taking cuts less than 25 μm.

*Polishing:*

a. With alumina slurry (1 μm), using a rapidly rotating wheel with a soft cloth and some soap addition. If the sample surface tarnishes, rinse with a solution of distilled water and 1 g of ammonium acetate per liter.
b. Polish In with alumina slurry and soap and etch with Cd m6 intermittently.
c. Electrolytic polishing with:
   Cd EP1 for Cd, Tl
   Cd EP2 for In
   Cd EP3 for Cd
d. Chemical polishing with Cd CP1 and Cd CP2 for Cd.

II. Soft solders: 84Cd-16Zn.

Alloys of Cd, In, and Tl. These alloys are considerably harder than pure metals.

*Grinding:*

a. Use a special file for light and soft metals.
b. Wet grinding on SiC papers down to 1200 grit. Coat papers with paraffin or stearin if necessary.
c. Ultramilling with hardmetal or diamond knife, this saves polishing; the sample must be very well cooled during the milling operation.

*Polishing:*

With 3, 1, and 0.25 μm diamond particle size on a cotton cloth and a short-nap synthetic cloth. Final polishing with $SiO_2$ suspension.

## Electrochemical Polishing Recipes

*Electrolytic Polishing*

| | | |
|---|---|---|
| Cd EP1 | 550 mL distilled water<br>450 mL phosphoric acid 85% | 30 min<br>2 V DC, Ni cathode |
| Cd EP2 | 670 mL methanol 99.8%<br>330 mL nitric acid 65% | 1–2 min<br>40–50 V DC<br>Stainless steel cathode<br>Do not store!<br>!!! See section 4.4 in the Appendix. |
| Cd EP3 | 200 mL distilled water<br>400 mL glycerol 87%<br>400 mL phosphoric acid 85% | 5–10 min<br>8–9 V DC<br>Suitable for polarized light examination |

*Chemical Polishing*

| | | |
|---|---|---|
| Cd CP1 | 25 mL distilled water<br>75 mL nitric acid 65% | 5–10 s |
| Cd CP2 | 20 mL ethanol 96%<br>10 mL nitric acid 65%<br>20 mL hydrogen peroxide 30% | 2 min<br>When mixing, always add the nitric acid to the ethanol.<br>!!! See section 4.4 in the Appendix. |

## Microetching

| No. | Material | Etchant | Conditions |
|---|---|---|---|
| Cd m1 | Pure Cd. Pure Tl<br>Cd alloys | 98 mL ethanol 96%<br>2 mL nitric acid 65%<br>(Nital reagent) | Seconds to minutes |
| Cd m2 | Cd-Ag alloys | 100 mL distilled water<br>2 mL sulfuric acid 95–97%<br>5 g chromium (VI) oxide | Seconds to minutes<br>!!! See section 4.4 in the Appendix. |
| Cd m3 | Pure Cd. Pure Tl<br>Soldering alloys with Cd | 100 mL distilled water<br>10 g chromium (VI) oxide | 1 to 10 min<br>!!! See section 4.4 in the Appendix. |

(continued)

## Microetching (continued)

| No. | Material | Etchant | Conditions |
|---|---|---|---|
| Cd m4 | Cd-Sn and Cd-Zn eutectics | 100 mL distilled water<br>25 mL hydrochloric acid 32%<br>8 g iron (III) chloride | Seconds to minutes |
| Cd m5 | Pure Cd, In, Tl<br>In-Sb and In-As alloys | 40 mL distilled water<br>10 mL hydrofluoric acid 40%<br>10 mL hydrogen peroxide 30% | 5–10 s<br>!!! See section 4.4 in the Appendix. |
| Cd m6 | Pure In<br>In-rich alloys | 100 mL ethanol 96%<br>5 mL hydrochloric acid 32%<br>1 g picric acid | Seconds to minutes<br>For Bi containing In: increase the amount of hydrochloric acid and ethanol.<br>!!! See section 4.4 in the Appendix. |
| Cd m7 | In-Sb alloys | 10 mL distilled water<br>50 mL hydrofluoric acid 40%<br>50 mL nitric acid 65% | 5–10 s<br>Etch no other metals in this solution.<br>!!! See section 4.4 in the Appendix. |

## Electrolytic Etching

| No. | Material | Etchant | Conditions |
|---|---|---|---|
| Cd EM1 | Pure Cd<br>Pure In | 100 mL distilled water<br>200 mL glycerol 87%<br>200 mL phosphoric acid 85% | 5–10 min<br>8–9 V DC<br>Cd cathode<br>Suited for examination in polarized light |

## *1.9.7 Cobalt*

Co

### Preparation

I. Pure cobalt.

Pure cobalt is tough and tends to form smear and deformation layers.

*Grinding:*

Wet grinding on SiC papers down to 1200 grit with light pressure.

*Polishing:*

a. With 6, 3, and 1 μm diamond particle size. Final polishing with 0.3 and 0.05 μm alumina slurry. If necessary, etch between polishing steps with 3% nitric acid in alcohol.
b. With 6 μm diamond particle size, then for 2 to 3 h on vibratory polisher charged with 0.3 μm alumina slurry or chromium oxide. Final polishing in a 10% aqueous $SiO_2$ suspension.
c. Electrolytic polishing with Co EP2.

II. Magnetic alloys:

Co-rich alloys containing more than 50 Co, e.g., Co-Ni, Co-Fe, Co-Pt, Co-V, Co-Sm.

Co alloys containing less than 50 Co, e.g., 25 to 30 Co with Al, Cu, Fe, Li, Ni (Alnico). 12 Co with Fe, Mo (Remalloy). 29 Co with Cu, Ni (Unico). Fe-Ni-Co alloys. Fe34Co with Ni, Al, Cu, Ti (Ticonal X).

Cobalt steels with 3 to 36 Co.

The preparation of these diverse Co alloys is not very difficult. The mechanical properties range from ductile to hard-brittle, and the preparation has to be adjusted accordingly.

*Grinding:*

Wet grinding on SiC papers down to 1200 grit.

*Polishing:*

a. For hard and brittle alloys polish with 6, 3, 1, and 0.25 μm diamond particle size. Final polishing with 0.05 μm alumina slurry.
b. For ductile alloys polish with 6 to 1 μm diamond particle size; this is followed by shock polishing with the electrolyte Co EP1 or Co EP5.
c. Electrolytic polishing: Alnico with Co EP2; intermetallic alloy with Co EP5. Co alloys tarnish easily and should only be rinsed with alcohol.

III. High-temperature-resistant Co-base alloys (superalloys):

Co with 18 to 35 Cr, 5W-Mo, and Zr, Ti, Al, Ta, B, Si, Mn. Cast and wrought alloys.

Co superalloys are hard and tough.

*Grinding:*

a. Wet grinding on SiC papers down to 1200 grit.
b. Grinding on SiC papers down to 320 grit, then lapping on a synthetic-bonded lapping disk with 6 μm diamond particle size.

*Polishing:*

a. Polish with 3 and 1 μm diamond particle size on a hard synthetic cloth. Final polish with 0.05 μm alumina on a hard silk cloth.
b. Electrolytic polishing with Co EP4.

IV. Stellite:

45 to 60 Co-Cr-W with 2 to 3 C, Si, Mn. Casting alloys.

These materials have excellent abrasion resistance and are very hard and brittle.

*Grinding:*

a. Wet grinding on SiC papers down to 320 grit, then lapping on a synthetic-bonded lapping disk with 6 μm diamond particle size.
b. Grinding on a diamond disk with 20 μm particle size.

*Polishing:*

Polish with 6 μm diamond particle size on a hard synthetic cloth. Final polish with 1 μm diamond particle size on a cotton cloth.

V. Hardmetals: see section 2.8.6 in Chapter 2.

**Electrochemical Polishing Recipes**

*Electrolytic Polishing*

| | | | |
|---|---|---|---|
| Co<br>EP1 | 600 mL<br>330 mL | methanol 99.8%<br>nitric acid 65% | 10–60 s<br>40–70 V DC<br>Stainless steel cathode. Do not store!<br>!!! See section 4.4 in the Appendix. |
| Co<br>EP2 | | Phosphoric acid 85% | 3–5 min. 1.5 V DC<br>Stainless steel cathode |
| Co<br>EP3 | 600 mL<br>400 mL | distilled water<br>phosphoric acid 85% | 1–15 min. 1–2 V DC<br>Stainless steel or Co cathode.<br>Rinsing will remove black deposit. |
| Co<br>EP4 | 900 mL<br>100 mL | distilled water<br>sulfuric acid 95–97% | Seconds to minutes. 6 or 4 V DC<br>Stainless steel cathode |
| Co<br>EP5 | 200 mL<br>700 mL<br>100 mL | perchloric acid 60%<br>ethanol 96%<br>butyl glycol 99% | Seconds to minutes. 30–60 V DC<br>Stainless steel cathode<br>!!! See section 4.4 in the Appendix. |

*Chemical Polishing*

| | | | |
|---|---|---|---|
| Co<br>CP1 | 40 mL<br>30 mL<br>5 mL | lactic acid 90%<br>hydrochloric acid 32%<br>nitric acid 65% | Seconds to minutes |
| Co<br>CP2 | 100 mL<br>5 g | distilled water<br>chromium (VI) oxide | Seconds to minutes<br>!!! See section 4.4 in the Appendix. |

(continued)

*Chemical Polishing (continued)*

| | | |
|---|---|---|
| Co CP3 | 50 mL glacial acetic acid<br>50 mL nitric acid 65% | Seconds to minutes |

## Macroetching

| No. | Material | Etchant | Conditions |
|---|---|---|---|
| Co M1 | Co-Cr alloys<br>Stellite | 50 mL distilled water<br>50 mL hydrochloric acid 32% | 30–60 min<br>Hot. Rinse with warm water. |
| Co M2 | Co-25Cr-10Ni-8W<br>Co-21Cr-20Ni<br>Co-3Cr-3Mo-1Nb<br>Stellite | 100 mL distilled water<br>10 mL nitric acid 65%<br>50 mL hydrochloric acid 32%<br>10 g iron (III) chloride | Swabbing<br>Rinse with warm water. |
| Co M3 | Co-Ni, Fe alloys<br>High-temperature-resistant alloys | 25 mL distilled water<br>50 mL hydrochloric acid 32%<br>25 mL nitric acid 65% | Seconds to minutes<br>!!! See section 4.4 in the Appendix. |

## Microetching

| No. | Material | Etchant | Conditions |
|---|---|---|---|
| Co m1 | Pure Co<br>Co-Fe alloys | 100 mL methanol 99.8%<br>1–50 mL nitric acid 65% | Seconds to minutes<br>Neutralize after use!<br>!!! See section 4.4 in the Appendix. |
| Co m2 | Pure and low-alloy Co<br>Co-B, Co-Ti, and Co-Mn alloys<br>WC-TiC-TaC-Co hardmetals<br>Grain-boundary etchant | 15 mL distilled water<br>15 mL glacial acetic acid<br>60 mL hydrochloric acid 32%<br>15 mL nitric acid 65% | 5–30 s<br>Mix and use only after 1 h.<br>!!! See section 4.4 in the Appendix. |
| Co m3 | Magnetic alloys<br>Co-Fe alloys | 100 mL distilled water<br>100 mL hydrochloric acid 32%<br>200 mL methanol 99.8%<br>5 mL nitric acid 65%<br>7 g iron (III) chloride<br>2 g copper (II) chloride | 10–15 s<br>Immersion or swab etching |
| Co m4 | Co-Pt alloys<br>WC-TiC-NbC-Co hardmetals | 75 mL hydrochloric acid 32%<br>25 mL nitric acid 65% | Up to 5 min<br>Use fresh mixture only!<br>!!! See section 4.4 in the Appendix. |
| Co m5 | Co-Ga, Co-Fe-V, and Co-Ni-V alloys | 100 mL distilled water<br>25 mL nitric acid 65% | 90 s |
| Co m6 | Co-Sm alloys<br>Color etchant | 100 mL distilled water<br>2 mL hydrochloric acid 32%<br>20 g potassium disulfite | 10 s |
| Co m7 | Co-Sm alloys<br>Grain-boundary etchant | 100 mL distilled water<br>8 g chromium (VI) oxide<br>2 g sodium sulfate | 5–10 s<br>Rinse in hot water and remove smudge layer with 20% aqueous chromium (VI) oxide solution.<br>!!! See section 4.4 in the Appendix. |

(continued)

## Microetching (continued)

| No. | Material | Etchant | Conditions |
|---|---|---|---|
| Co m8 | Co-Sm alloys | 100 mL distilled water<br>1 mL glacial acetic acid<br>1 mL nitric acid 65% | Several s |
| Co m9 | Co superalloys | 200 mL hydrochloric acid 32%<br>5 mL nitric acid 65%<br>65 g iron (III) chloride | Seconds to minutes |
| Co m10 | Co superalloys | 50 mL distilled water<br>50 mL hydrochloric acid 32%<br>10 g copper (II) sulfate (Marble's etch) | Seconds to minutes<br>Immersion or swabbing<br>Add few drops of sulfuric acid to activate the etchant. |
| Co m11 | Co borides | 30 mL distilled water<br>10 mL hydrochloric acid 32%<br>10 mL nitric acid 65% | Seconds to minutes<br>!!! See section 4.4 in the Appendix. |
| Co m12 | Co silicide | 100 mL distilled water<br>15 g chromium (VI) oxide | Seconds to minutes<br>Add few drops of hydrochloric acid before use.<br>!!! See section 4.4 in the Appendix. |
| Co m13 | Co and Co-Al alloys | 25 mL distilled water<br>50 mL hydrochloric acid 32%<br>15 g iron (III) chloride<br>3 g ammonium tetrachlorocuprate (II)<br>(Adler's reagent) | Seconds to minutes |
| Co m14 | Co-base superalloys<br>Color etchant<br>Carbides remain white. | 50 mL distilled water<br>50 mL hydrochloric acid 32%<br>2 g potassium disulfite | Seconds to minutes |
| Co m15 | Pure Co<br>Color etchant | 98 mL distilled water<br>2 mL hydrofluoric acid 40%, boiling, add molybdic acid to saturate | Seconds to minutes<br>RT<br>!!! See section 4.4 in the Appendix. |
| Co m16 | Co-base casting material and rolling stock<br>Stellite<br>Color etchant | 100 mL stock solution Beraha III<br>1 g potassium disulfite<br><br>**Stock solution Beraha III:**<br>600 mL distilled water<br>400 mL hydrochloric acid 32%<br>50 mL ammonium hydrogen difluoride | 30 s to 5 min<br>Wet etching<br>Keep only for 1–2 h.<br><br>Stock solution must be stored in plastic bottle!<br>!!! See section 4.4 in the Appendix. |

## Electrolytic Etching

| No. | Material | Etchant | Conditions |
|---|---|---|---|
| Co Em1 | Pure Co<br>Co-Al alloys | 100 m distilled water<br>5 mL hydrochloric acid 32%<br>10 g iron (III) chloride | Several s<br>6 V DC<br>Stainless steel cathode |

(continued)

## Electrolytic Etching (continued)

| No. | Material | Etchant | Conditions |
|---|---|---|---|
| Co<br>Em2 | Stellites up to 70% Co<br>Co-base superalloys | 100 mL distilled water<br>5–10 mL hydrochloric acid 32%<br>2–10 g chromium (VI) oxide | 2–20 s<br>3 V DC<br>Stainless steel cathode<br>!!! See section 4.4 in the Appendix. |
| Co<br>Em3 | Pure Co<br>Co-base superalloys | 100 mL distilled water<br>5–10 mL hydrochloric acid 32% | 2–10 s<br>3 V DC<br>Graphite cathode |
| Co<br>Em4 | Co-base abrasion-resistant alloys and tool materials<br>Superalloys | 100 mL hydrochloric acid 32%<br>5 mL hydrogen peroxide 30% | 3–5 s<br>4 V DC<br>Stainless steel cathode |

## Physical Etching

The carbides in Co-base superalloys and stellites can be identified by reactive sputtering with a Fe cathode in an $O_2$ atmosphere. Intermetallic phases in Co-Sm alloys can be distinguished by reactive sputtering with an Au cathode in an $O_2$ atmosphere. Vaporizing of ZnSe is recommended to improve the contrast of the Kerr effect in magnetic alloys.

Cr Mo Nb Re Ta V W

### *1.9.8 Refractory Metals (Chromium, Molybdenum, Niobium, Rhenium, Tantalum, Vanadium, and Tungsten)*

## Preparation

I. The pure metals have a high melting point and are corrosion resistant. Their application is high-temperature technology, tool industry, lighting engineering, semiconductor, Cr galvanizing process, and nuclear, reactor, and radiation technology; other applications are in magnets with high magnetic fields because of the superconductivity of Nb alloys and in the medical field for implants.

Pure Nb, Ta, V, and Re are very soft and ductile, while Cr, Mo, and W are rather brittle at room temperature.

Because of their poor cutting ability, mechanical grinding and polishing is quite limited. Recommended are combined methods, e.g., chemical-mechanical polishing or attack polishing with intermediate etching steps. The intermediate etching between polishing steps will help to remove the deformation layers.

*Grinding:*

Wet grinding on SiC papers down to the smallest grit with little pressure.

*Polishing:*

a. Polish with 6 to 3 μm diamond particle size using alcohol-based lubricants on cotton cloth, then with 1 μm diamond on short-nap velvet cloth. Etching between polishing steps (intermediate etching) is recommended to remove the deformation layer. Highly diluted Cr EP2 and Cr EP5 can be used for Cr, Mo, Nb, Re, Ta, V, and W.
   Use Cr m9 for Mo, Re, Ta, and W. For Mo and W: Cr m12 and for Nb and Ta: Cr m18.
   Final polishing on a vibratory polisher or in a polishing tank, where the sample is automatically moved, using a long-nap, acid-resistant polishing cloth. Use alumina slurry, $SiO_2$ suspension, or one of the solutions Cr CM1 to Cr CM7.
   Difficult-to-remove scratches can be removed by shock polishing with the electrolytes Cr EP1 to Cr EP5.
b. Following grinding, use vibratory polishing on a soft, short-nap cloth with alumina suspension for 10 to 20 h.
c. Following grinding, chemical-mechanical polish on a vibratory polisher or in a polisher with tank and automatic sample movement on a soft and acid-resistant synthetic cloth with one of the solutions of Cr CM1 to Cr CM7.
d. Polish with 3 μm diamond particle size and then electrolytically with an electrolyte from Cr EP1 to Cr EP5.
e. Polish with 3 μm diamond particle size followed by electrolytic lapping with an electrolyte from Cr EW1 to Cr EW4.

II. Technically pure metals with nonmetallic additions. Interstitial impurities of N, O, and C increase the hardness and tend to embrittle the material. The hard, nonmetallic precipitates in the matrix cause relief problems and possible breakouts.

*Grinding:*

Wet grinding on SiC papers down to 800 grit, coat papers with 1200 and 2400 grit with stearin and grind without lubricant.

*Polishing:*

Immediately after grinding, polish with a vibratory polisher on a short-nap velvet cloth with a 5 μm alumina suspension.

Final polish with $SiO_2$ suspension.

III. Refractory metal alloys. Additions of $ThO_2$, Al, Cu, Fe, Re, Sn, Ti, Zr, and Ni-Fe or Ni-Cu alloys increase the strength, heat resistance, electrical conductivity, thermal conductivity, corrosion resistance, and formability.

*Grinding:*

a. As for I.a.
b. As for II.

*Polishing:*

As for I.a to I.e.

The preparation times of the individual steps are shorter for alloys than for pure metals.

## Electrochemical Polishing Recipes

*Electrolytic Polishing*

| | | | |
|---|---|---|---|
| Cr<br>EP1 | Cr | 210 mL distilled water<br>640 mL phosphoric acid 85%<br>150 mL sulfuric acid 95–97% | Seconds to minutes<br>22–120 °C<br>1.5 A/cm$^2$. 18 V DC<br>Stainless steel cathode<br>!!! See section 4.4 in the Appendix. |
| Cr<br>EP2 | Cr<br>W | 1000 mL distilled water<br>100 g potassium hydroxide | 10–25 min<br>6 V DC<br>Stainless steel or graphite cathode<br>!!! See section 4.4 in the Appendix. |
| Cr<br>EP3 | Mo | 875 mL methanol 99.8% or distilled water<br>125 mL sulfuric acid 95–97% | 30 s to 2 min<br>6–18 V DC<br>Stainless steel cathode<br>!!! See section 4.4 in the Appendix. |
| Cr<br>EP4 | Mo<br>Ta<br>Nb | 940 mL methanol 99.8%<br>50 mL sulfuric acid 95–97%<br>10 mL hydrofluoric acid 40% | 10–20 s<br>50–60 V DC<br>Stainless steel cathode<br>!!! See section 4.4 in the Appendix. |

(continued)

*Electrolytic Polishing (continued)*

| | | | |
|---|---|---|---|
| Cr<br>EP5 | Mo<br>Re alloys<br>Re-Mo alloys | 100 mL distilled water<br>400 mL glacial acetic acid<br>100 mL chromium (VI) oxide | 30 s to 3 min<br>50 V DC<br>Stainless steel cathode<br>!!! See section 4.4 in the Appendix. |

*Electrolytic Lapping*

| | | | |
|---|---|---|---|
| Cr<br>EW1 | Re | 100 mL distilled water with 0.05 μm alumina<br>10 mL saturated aqueous potassium chloride solution<br>(Variable concentration) | 2 min<br>0.1 $A/cm^2$ |
| Cr<br>EW2 | W | 100 mL distilled water<br>2 g sodium hydroxide | 1–10 min<br>10–20 V DC<br>Stainless steel cathode. Short-nap velvet cloth |
| Cr<br>EW3 | Mo, Re, W alloys | 30 mL distilled water with 0.05 μm alumina<br>1 g potassium ferricyanide | 5–10 min<br>40 V DC<br>Stainless steel cathode. Short-nap velvet cloth |
| Cr<br>EW4 | Nb, M, W | 300 mL distilled water with 0.05 μm alumina<br>10 mL hydrogen peroxide 30% | 20 min<br>1–10 $mA/m^2$<br>Stainless steel cathode |

*Chemical Polishing*

| | | | |
|---|---|---|---|
| Cr<br>CP1 | Nb, V, Ta | 120 mL distilled water<br>30 mL hydrochloric acid 32%<br>16 mL hydrofluoric acid 40%<br>6 g iron (III) chloride | 1 min for V<br>2 min for Nb<br>3 min for Ta<br>!!! See section 4.4 in the Appendix. |

(continued)

*Chemical Polishing (continued)*

| | | | |
|---|---|---|---|
| Cr CP2 | Nb, V, Ta | 30 mL distilled water<br>30 mL nitric acid 65%<br>30 mL hydrochloric acid 32%<br>15 mL hydrofluoric acid 40% | Seconds to minutes<br>!!! See section 4.4 in the Appendix. |
| Cr CP3 | Nb, Ta | 50 mL sulfuric acid 95–97%<br>20 mL nitric acid 65%<br>20 mL hydrofluoric acid 40% | 5–10 s<br>!!! See section 4.4 in the Appendix. |

*Chemical-Mechanical Polishing*

| | | | |
|---|---|---|---|
| Cr CM1 | Cr | 150 mL distilled water with 0.05 μm alumina<br>15 mL glacial acetic acid | 5–10 min<br>Acid-resistant cloth |
| Cr CM2 | Ta | 78 mL distilled water with 0.05 μm alumina<br>15 mL hydrogen peroxide 30%<br>12 g potassium hydroxide | Seconds to minutes<br>Acid-resistant cloth |
| Cr CM3 | W | 100 mL distilled water<br>100 mL ammonium hydroxide 25%<br>100 mL hydrogen peroxide 30% | Seconds to minutes<br>Add hydrogen peroxide last.<br>!!! See section 4.4 in the Appendix. |
| Cr CM4 | Cr, Mo, Nb, Ta, V, W, and their alloys | 35 mL distilled water with 0.05 μm aluminum oxide<br>5 mL aqueous solution of chromium (VI) oxide 20% | Seconds to minutes<br>!!! See section 4.4 in the Appendix. |
| Cr CM5 | Mo, Nb, Ta, W | 100 mL distilled water with 0.05 μm alumina<br>5 g sodium hydroxide | Seconds to minutes<br>Polish down to 1 μm diamond. |
| Cr CM6 | Mo, W | 150 mL distilled water with 0.05 μm alumina<br>3.5 g potassium ferricyanide<br>1 g sodium hydroxide | Seconds to minutes<br>Use chemically resistant synthetic cloth |

(continued)

*Chemical-Mechanical Polishing (continued)*

| | | | |
|---|---|---|---|
| Cr CM7 | $Nb/Nb_2O_5$ W-Ni-Fe | 300 mL distilled water with 0.05 μm alumina<br>20 mL hydrogen peroxide 30% | Seconds to minutes<br>Use chemically resistant synthetic cloth. |

## Macroetching

Surface prefinish: Wet grinding on SiC papers down to 1000 grit.

| No. | Material | Etchant | Conditions |
|---|---|---|---|
| Cr M1 | Cr | 90 mL distilled water<br>10 mL sulfuric acid 95–97% | 2–5 min<br>Boiling |
| Cr M2 | Mo, Nb, Ta, V, W, and their alloys | 30 mL hydrochloric acid 32%<br>15 mL nitric acid 65%<br>30 mL hydrofluoric acid 40% | Seconds to minutes<br>!!! See section 4.4 in the Appendix. |
| Cr M3 | Mo, V, and W | 75 mL distilled water<br>35 mL nitric acid 65%<br>15 mL hydrofluoric acid 40% | 10–20 min<br>!!! See section 4.4 in the Appendix. |

## Microetching

| No. | Material | Etchant | Conditions |
|---|---|---|---|
| Cr m1 | Mo<br>Color etchant | 75 mL ethanol 96%<br>40–60 mL aqueous iron (III) chloride $6 H_2O$ solution (1300 g/L, 50 °C)<br>25 mL hydrochloric acid 32%<br>(Hasson's reagent) | 4 min at RT |
| Cr m2 | Mo | 70 mL distilled water<br>20 mL hydrogen peroxide 30%<br>10 mL sulfuric acid 95–97% | 2 min<br>Immersion etching<br>!!! See section 4.4 in the Appendix. |
| Cr m3 | Mo<br>Etch figures | 30 mL methanol 99.8%<br>15 mL nitric acid 65%<br>6 mL sulfuric acid 95–97% | Seconds to minutes<br>!!! See section 4.4 in the Appendix. |
| Cr m4 | W<br>Color etchant | 94 mL aqueous solution of hydrochloric acid 10%<br>20 g chromium (VI) oxide | 15 min at 55 °C, then 10 min, and an additional 10 min<br>Pre-etching with a grain-boundary etchant produces better results.<br>!!! See section 4.4 in the Appendix. |
| Cr m5 | Cr and Cr-base alloys | 20 mL nitric acid 65%<br>60 mL hydrochloric acid 32% | 5–60 s<br>Do not store!<br>!!! See section 4.4 in the Appendix. |
| Cr m6 | W and W-base alloys | 100 mL distilled water<br>1 mL hydrogen peroxide 30%<br>(Concentration variable) | Boiling for 30–90 s |
| Cr m7 | Cr and Nb and their alloys | 20 mL nitric acid 65%<br>60 mL hydrofluoric acid 40% | Approximately 10 s<br>!!! See section 4.4 in the Appendix. |

(continued)

## Microetching (continued)

| No. | Material | Etchant | Conditions |
|---|---|---|---|
| Cr m8 | Cr, Mo, Mo-Cr alloys (up to 80% Cr)<br>Mo-Fe alloys and W-base alloys<br>Mo-Re alloys<br>Re and Re-base alloys | (a) 100 mL distilled water<br>10 g potassium hydroxide<br>(b) 100 mL distilled water<br>10 g potassium ferricyanide | 15–60 s<br>Mix even parts of (a) and (b).<br>Use fresh mixture only!<br>Sodium hydroxide and sodium ferricyanide may be used for Mo and W. |
| Cr m9 | Mo and Mo-Ni alloys<br>W and W alloys<br>Nb and Nb alloys | 50 (70) mL distilled water<br>50 (20) mL hydrogen peroxide 30%<br>50 (10) mL ammonium hydroxide solution 32% | Seconds to minutes for Mo and Mo-Ni alloys<br>10 min boiling for W and W alloys<br>Data in parentheses are for Nb and Nb alloys. |
| Cr m10 | Nb and Nb-base alloys<br>Ta and Mo and their base alloys<br>Mo-Hf alloys<br>W and Re<br>Re-Hf alloys<br>V and V-base alloys | 10 (10) mL hydrofluoric acid 40%<br>30 (10) mL nitric acid 65%<br>60 (30) mL lactic acid 90% | 15–20 s<br>Do not store!<br>Data in parentheses are for Ta and Nb alloys.<br>!!! See section 4.4 in the Appendix. |
| Cr m11 | W, Ta, Nb, and their base alloys<br>Cr and Cr silicide<br>Re silicide. W-Th alloys<br>$Nb_3Sn$ | 50 (50) mL distilled water<br>50 (25) mL nitric acid 65%<br>50 (5) mL hydrofluoric acid 40%<br>(Concentration variable) | Seconds to minutes<br>!!! See section 4.4 in the Appendix. |
| Cr m12 | Mo-Al and Mo-Co alloys | 50 mL methanol 99.8%<br>50 mL glacial acetic acid<br>50 mL nitric acid 65% | Seconds to minutes<br>!!! See section 4.4 in the Appendix. |
| Cr m13 | Nb and Nb alloys<br>Ta and Ta-O alloys (industrial alloys)<br>Etch figures | 20 mL distilled water<br>20 mL sulfuric acid 95–97%<br>20 mL hydrofluoric acid 40%<br>2 mL hydrogen peroxide 30% | Seconds to minutes<br>Cool during mixing!<br>Neutralize with 25% aqueous sodium hydroxide.<br>!!! See section 4.4 in the Appendix. |
| Cr m14 | Nb-Zr alloys | 25 mL ethanol 96%<br>50 mL hydrogen peroxide 30%<br>25 mL nitric acid 65%<br>1 mL hydrofluoric acid 40% | Seconds to minutes<br>Do not store!<br>!!! See section 4.4 in the Appendix. |
| Cr m15 | Nb and Nb-base alloys<br>Mo and Mo alloys<br>Ta and Ta-base alloys<br>Nb-Cr alloys<br>$Nb_3Sn$ | 50 (0) mL distilled water<br>20 (20) mL hydrofluoric acid 40%<br>10 (20) mL nitric acid 65%<br>15 (50) mL sulfuric acid 95–97%<br>(Concentration variable) | Seconds to minutes<br>!!! See section 4.4 in the Appendix. |
| Cr m16 | Nb-base alloys<br>Ta-base alloys<br>W-base alloys | 50 mL glacial acetic acid<br>20 mL nitric acid 65%<br>5 mL hydrofluoric acid 40%<br>(Concentration variable) | 10–30 s<br>!!! See section 4.4 in the Appendix. |
| Cr m17 | Ta and Ta-base alloys<br>Nb-Re alloys | 50 mL distilled water<br>50 (20) mL hydrofluoric acid 40%<br>(Concentration variable) | 10 s<br>Data in parentheses are for Nb-Re alloys.<br>!!! See section 4.4 in the Appendix. |

(continued)

## Microetching (continued)

| No. | Material | Etchant | Conditions |
|---|---|---|---|
| Cr m18 | W-Co alloys with 10–70 W | 10 mL hydrochloric acid 32%<br>10 mL aqueous solution of hydrogen peroxide 3% | Seconds to minutes |
| Cr m19 | W-Co alloys (eutectic)<br>W phase turns black. | 100 mL distilled water<br>25 g sodium hydroxide<br>2 g picric acid | Boiling for 15 s<br>!!! See section 4.4 in the Appendix. |

## Electrolytic Etching

| No. | Material | Etchant | Conditions |
|---|---|---|---|
| Cr Em1 | Cr | 95 mL distilled water<br>5 mL sulfuric acid 95–97% | Seconds to minutes<br>2–3 V DC<br>5–10 V DC (for Cr alloys) |
| Cr Em2 | Mo, W | 100 mL methanol 99.8%<br>20 mL sulfuric acid 95–97% | Seconds to minutes<br>5 V DC. 100 mA/cm$^2$<br>Stainless steel cathode<br>When etching W, use electrolyte only for W. |
| Cr Em3 | Mo, Nb | 65 mL distilled water<br>17 mL nitric acid 65%<br>17 mL hydrofluoric acid 40% | Seconds to minutes<br>12–50 V DC<br>Pt cathode<br>!!! See section 4.4 in the Appendix. |
| Cr Em4 | Cr and Cr-base alloys<br>Fe-Cr alloys | 95 mL glacial acetic acid<br>5 mL perchloric acid 60% | 15 s<br>30–50 V DC<br>Stainless steel or Pt cathode<br>!!! See section 4.4 in the Appendix. |
| Cr Em5 | Cr and Cr-base alloys<br>Cr-Fe alloys<br>Mo (grain-contrast etchant)<br>Re and Re-base alloys<br>V and V-base alloys | 100 mL distilled water<br>10 g oxalic acid | 2–5 s<br>1 min for Re and Re-base alloys<br>6 V DC<br>Stainless steel cathode<br>!!! See section 4.4 in the Appendix. |
| Cr Em6 | High-Cr-containing alloys<br>Cr silicide<br>Galvanic Cr layers<br>Mo-Cr-Fe alloys<br>U-Nb alloys<br>V and V-base alloys<br>Ferrovanadium | 95 mL distilled water or ethanol 96%<br>5 mL hydrochloric acid 32% | Seconds to minutes<br>5–10 V DC<br>Stainless steel cathode |
| Cr Em7 | Mo and Mo-base alloys<br>W and W-base alloys | 100 mL methanol 99.8%<br>5 mL sulfuric acid 95–97%<br>1 mL hydrofluoric acid 40% | 10–20 s<br>50–60 V DC<br>Stainless steel cathode<br>!!! See section 4.4 in the Appendix. |
| Cr Em8 | Mo and Mo-base alloys<br>Ta and Ta-base alloys<br>V and V-base alloys | 75 mL methanol 99.8%<br>10 mL sulfuric acid 95–97%<br>25 mL hydrochloric acid 32% | 30 s<br>30 V DC<br>Stainless steel cathode<br>!!! See section 4.4 in the Appendix. |
| Cr Em9 | Mo and its alloys<br>Ta and its alloys<br>W and W-base alloys | 100 mL distilled water<br>10 g sodium hydroxide | Seconds to minutes<br>1.5–6 V DC<br>Stainless steel cathode |

## Anodizing

| No. | Materials | Etchant | Conditions |
|---|---|---|---|
| Cr An1 | Nb<br>NbGa<br>$Nb_3Si$ | 50 mL ethanol 96%<br>10 mL distilled water<br>1 mL phosphoric acid 85% | 1 min<br>80 V DC |
| Cr An2 | $Nb_3Sn$ | 100 mL distilled water<br>1 mL phosphoric acid 85% | 1 min<br>26 V DC |
| Cr An3 | Ta and Ta alloys<br>Color etchant | 100 mL distilled water<br>10 g sodium sulfite | Seconds to minutes<br>Up to 240 V DC (8 mA/cm$^2$). Heat sample after anodizing for 5 min to 800 °C under vacuum, then etch for 3 s in hydrofluoric acid 40%.<br>!!! See section 4.4 in the Appendix. |
| Cr An4 | Ta and Ta alloys | 35 mL distilled water<br>60 mL ethanol 96%<br>20 mL glycerol 87%<br>10 mL lactic acid 90%<br>5 mL phosphoric acid 85%<br>2 g citric acid (powdered) | 10–60 s<br>20–140 V DC |
| Cr An5 | Ta<br>W | 998 mL distilled water<br>2 mL phosphoric acid 85% | 4 min (5–30 s)<br>240 V DC (60–80 V DC)<br>Use conditions in parentheses for W. |
| Cr An6 | Pure V<br>V-Al, V-Ga, V-In, V-Si alloys | 20 mL distilled water<br>90 mL ethanol 96%<br>20 mL glycerol 87%<br>1.5 mL phosphoric acid 85% | 60–100 s<br>35 V DC<br>!!! See section 4.4 in the Appendix. |

Cu

### *1.9.9 Copper*

#### Preparation

I. Pure copper.

Copper containing oxygen, such as electrolytic Cu 99.9, oxygen-free Cu, such as OFHC-Cu 99.9, are used for tubes and applications in the electrical and electronics industry.

II. Low-alloy Cu materials:

They contain alloying elements of up to 5% by weight and are used mainly in the electronics and instrumentation technology.

Nonhardenable alloys: Cu-Ag, Cu-As-P, Cu-Cd, Cu-Cd-Sn, Cu-Mg, Cu-Mn, Cu-Si-Mn, Cu-Te-P, Cu-Fe, and Cu-Zn.

Hardenable alloys: Cu-Be, Cu-Ca-Be, Cu-Ni-Be, Cu-Cr, Cu-Cr-Zr,and Cu-Ni-Si.

III. Cu alloys:

Brasses: Cu with up to 50 Zn, with or without Pb additions. Wrought and foundry alloys. Special brasses containing Al, Fe, Mn, Ni, Si, and/or Sn are used for drawn parts, fittings, screws, bearings, etc., parts for the electrical industry, watchmaking and jewelry industry.

Bronzes: Sn bronzes with up to 20 Sn are used for bells, fittings, etc. Al bronzes with 4 to 11 Al and additions of Ni, Mn, and Fe have their applications as corrosion-resistant materials.

Pb bronzes with up to 28 Pb with Sn, Ni, and Zn are used as bearing materials.

Red bronzes: Cu-Sn-Zn alloys, also with Pb are used as bearing materials.

Be bronzes with up to 2.5 Be are used in spring materials, spark free tools, etc.

Cu-Ni alloys: Cu-Ni with 5 to 44 Ni. Foundry and wrought alloys are used for tubes, instrumentation, coins.

Alpaca: 10 to 25 Ni and 15 to 42 Zn; with additions of Pb, Mn, Sn. Flatware, kitchen ware, electrical industry, electronics, fine mechanics, structural parts.

The Cu materials groups I to III can all be prepared using the same method. It is important to use low pressure when grinding and polishing pure Cu to avoid deformation layers and the embedding of abrasive grains into the sample surface.

*Grinding:*

a. Wet grinding on SiC papers down to 800 grit for macroetching.
b. Ultramilling with a hardmetal or diamond knife for materials of groups I and II, with the exception of hardenable alloys; polishing will not be necessary.
c. Wet grinding on SiC papers down to 1200 grit, then wet grinding on microcut paper.

*Polishing:*

a. Polish after using microcut paper with 3 and 1 μm diamond particle size on a cotton cloth using an oil-based lubricant.
   Final polishing with 0.05 μm alumina slurry on a long-nap cloth adding a few drops of 25% ammonium hydroxide solution. The added amount must not etch the samples.
   Or: final polishing with $SiO_2$ suspension.
   Or: etch polishing with 0.05 μm alumina slurry and a 10% aqueous iron (III) nitrate solution. This will attack the grain boundaries slightly. Excellent for pure Cu.

b. For difficult-to-prepare samples etch between polishing steps with 10% aqueous iron (III) nitrate solution.
c. After polishing with 1 μm diamond particle size use shock polishing with Cu EP1.
d. Electrolytic polishing with electrolyte:
Cu EP1: Cu. alpha, beta, and alpha/beta brasses without Pb additions. Al, Si, and P bronze, Cu with less than 3 Be, Fe, Pb, or Cr
Cu EP2: Cu. Cu-Sn alloys with up to 6 Sn
Cu EP3: Al and Be bronze
Cu EP4: Pure Cu, Cu-Be, Cu-Cr, Cu-Al-Fe
Cu EP5: Cu-Ni, Cu-Sn, Cu-Ti, Cu-Co-Be
Cu EP6: Cu-Co, Cu-Mn, Cu-Zr, Cu-Cr-Zr
e. Electrolytic lapping with Cu EW1: High-purity Cu, rolled Cu, 80Ag-20Cu
f. Chemical polishing:
Cu CP1: Cu. alpha brass
Cu CP2: Cu
Cu CP3: Cu-Al (Si) and Cu-Be alloys
Cu CP4: All Cu types and Cu alloys

**Electrochemical Polishing Recipes**

*Electrolytic Polishing*

| | | |
|---|---|---|
| Cu EP1 | 300 mL distilled water<br>700 mL phosphoric acid 85% | 5–15 min<br>1.5–2 V DC<br>Cu cathode<br>Rinse in a 20% aqueous phosphoric acid solution after polishing. |
| Cu EP2 | 300 mL distilled water<br>670 mL phosphoric acid 85%<br>100 mL sulfuric acid 95–97% | 15 min<br>2 V DC<br>Cu cathode |
| Cu EP3 | 100 mL distilled water<br>1 g chromium (VI) oxide | 3–6 s<br>6 V DC<br>Al cathode<br>!!! See section 4.4 in the Appendix. |
| Cu EP4 | 594 mL methanol 99.8%<br>406 mL phosphoric acid 85% | 10 s. 15–25 V DC. 2.0–2.8 A/cm$^2$<br>Stainless steel cathode |

(continued)

*Electrolytic Polishing (continued)*

| | | |
|---|---|---|
| Cu<br>EP5 | 500 mL distilled water<br>100 mL ethylene glycol<br>400 mL phosphoric acid 85% | 5 s. 8–15 V DC. 2.0–3.0 A/cm$^2$<br>Stainless steel cathode |
| Cu<br>EP6 | 500 mL distilled water<br>100 mL ethylene glycol<br>400 mL phosphoric acid 85%<br>20 g ammonium acetate | 5 s. 8–12 V DC. 2.0–3.4 A/cm$^2$<br>Stainless steel cathode |

*Electrolytic Lapping*

| | | |
|---|---|---|
| Cu<br>EW1 | 885 mL distilled water<br>12 g sodium thiosulfate<br>2 g thiourea<br>1 g copper (II) nitrate | 5–7 min<br>High-purity Cu: 5 mA/0.4 cm$^2$<br>Rolled Cu: 8–9 mA/0.64 cm$^2$<br>80Ag-20Cu: 10 mA/0.8 cm$^2$ |

*Chemical Polishing*

| | | |
|---|---|---|
| Cu<br>CP1 | 25 mL phosphoric acid 85%<br>25 mL glacial acetic acid<br>25 mL nitric acid 65%<br>(0.5 mL hydrochloric acid 32%) | 2–4 min<br>55–80 °C<br>Agitate the solution. |
| Cu<br>CP2 | 65 mL glacial acetic acid<br>27 mL phosphoric acid 85%<br>6 mL nitric acid 65% | 1 min<br>60 °C<br>Grind sample down to 1200 grit. |
| Cu<br>CP3 | 30 mL nitric acid 65%<br>10 mL hydrochloric acid 32%<br>10 mL phosphoric acid 85%<br>50 mL glacial acetic acid<br>(Concentration variable) | 1–2 min<br>70–80 °C<br>Agitate the sample. |
| Cu<br>CP4 | 35 mL distilled water<br>40 mL nitric acid 65%<br>25 g chromium (VI) oxide<br>(Concentration variable) | Remove oxide film with 10% aqueous hydrofluoric acid.<br>!!! See section 4.4 in the Appendix. |

## Macroetching

| No. | Material | Etchant | | Conditions |
|---|---|---|---|---|
| Cu M1 | Pure Cu<br>Dendrite formation in alpha alloys<br>All brass types. Al bronzes<br>Grain-contrast etchant | 120 mL<br>30 mL<br>10 g | distilled water or ethanol 96%<br>hydrochloric acid 32%<br>iron (III) chloride<br>(Concentration variable) | A few min |
| Cu M2 | Pure Cu<br>All brasses<br>Grains and cracks | 90 mL<br>10–60 mL | distilled water<br>nitric acid 65% | A few min |
| Cu M3 | Pure Cu<br>For all Cu types and Cu alloys<br>Deep etchant | 50 mL<br>50 mL<br>5 g | distilled water<br>nitric acid 65%<br>silver nitrate | Seconds to minutes |
| Cu M4 | Pure Cu<br>Brasses<br>Grain-contrast etchant | 50 mL<br>50 mL | distilled water<br>nitric acid 65% | Seconds to minutes |
| Cu M5 | Brasses<br>Co-containing brass | 100 mL<br>25 g | distilled water<br>ammonium persulfate | Seconds to minutes |
| Cu M6 | Si-containing brasses and bronzes | 100 mL<br>50 mL<br>8 mL<br>40 g<br>7.5 g | distilled water<br>nitric acid 65%<br>sulfuric acid 95–97%<br>chromium (VI) oxide<br>ammonium chloride | Seconds to minutes<br>!!! See section 4.4 in the Appendix. |
| Cu M7 | Brasses<br>To reveal stresses<br>Stressed parts will crack after a while. | (a) 100 mL<br>1 g<br>(b) 100 mL<br>1 mL | distilled water<br>mercury (II) nitrate<br>distilled water<br>nitric acid 65% | Seconds to minutes<br>Mix (a) and (b), 1:1<br>The time the sample takes to crack is a measure for the amount of stresses present.<br>!!! See section 4.4 in the Appendix. |

## Microetching

| No. | Material | Etchant | | Conditions |
|---|---|---|---|---|
| Cu m1 | Pure Cu. Brasses. Bronzes<br>Al bronzes<br>Cu-Ni and Cu-Ag alloys. Alpaca | 100 mL<br>10 g | distilled water<br>ammonium persulfate (if necessary with 10 mL hydrochloric acid 32%) | Seconds to minutes<br>Heat if necessary.<br>Use fresh only! |
| Cu m2 | Pure Cu. Brasses. Alpaca<br>Bronzes<br>Grain-boundary etchant | | Saturated aqueous chromium (VI) oxide solution | 5–30 s<br>!!! See section 4.4 in the Appendix. |
| Cu m3 | Pure Cu with/without oxide and sulfide inclusions | 100 mL<br>100 mL<br>10 g | distilled water<br>ethanol 96%<br>iron (III) nitrate | Etch polishing without polishing compound on a short-nap velvet cloth. |
| Cu m4 | Pure Cu<br>Alpha/beta brass<br>Special brasses. Leaded bronze. Al bronze. Alpaca. Tin bronze | 120 mL<br>10 g | distilled water<br>ammonium tetrachlorocuprate (II)<br>Add ammonium hydroxide solution 25% until the precipitate is dissolved. | 5–60 s<br>If the solution is too concentrated, dilute with distilled water.<br>Use fresh only! |
| Cu m5 | Cu alloys. Brasses (discolors the beta phase). Cu-Be alloys<br>Al bronzes with eutectoid. Alpaca | 100–120 mL<br>20–50 mL<br>5–10 g | distilled water or ethanol 96%<br>hydrochloric acid 32%<br>iron (III) chloride | Seconds to minutes<br>If necessary re-etch with Cu m6. |

(continued)

## Microetching (continued)

| No. | Material | Etchant | | Conditions |
|---|---|---|---|---|
| Cu m6 | Brasses. Alpha bronze. Alpaca<br>Cu-Be, Cu-Cr, Cu-Mn, Cu-Ni, and Cu-Si alloys | 80 mL<br>5 mL<br>10 g | distilled water<br>sulfuric acid 95–97%<br>sodium or potassium dichromate solution | 3–30 s<br>Add 2 drops of hydrochloric acid 32%, just before use. |
| Cu m7 | For most Cu materials and Cu alloys. Alpha brass<br>Cu-Ag solder layers. Mn, Be, P, and AlSi bronze<br><br>Grain-boundary etchant<br><br>Grain-contrast etchant | 25 mL<br>25 mL<br>5–25 mL | distilled water<br>ammonium hydroxide solution 25%<br>aqueous hydrogen peroxide solution 3%<br><br>Add small amount of 3% hydrogen peroxide.<br>Add large amount of 3% hydrogen peroxide. | Seconds to minutes<br>Use fresh only!<br>Approximately 1–5 mL of aqueous potassium hydroxide solution 20% may be added. |
| Cu m8 | For most Cu materials and Cu alloys<br>Brasses (deformation bands) | 50 mL<br>50 mL | distilled water<br>nitric acid 65%<br>(Concentration variable) | Seconds to minutes |
| Cu m9 | Al bronzes (that are difficult to etch) | 100 mL<br>13 mL<br>6.6 mL | distilled water<br>nitric acid 65%<br>hydrofluoric acid 40% | Several min<br>!!! See section 4.4 in the Appendix. |
| Cu m10 | Al bronzes (only the gamma phase is attacked) | 100 mL<br>10 g | distilled water<br>sodium thiosulfate | 90–150 s<br>Sample must be agitated. |
| Cu m11 | Cu and Cu alloys<br>Be bronzes | 100 mL<br>1 mL<br>3 g | distilled water<br>ammonium hydroxide solution 25%<br>ammonium persulfate | Seconds to minutes |
| Cu m12 | Alloyed Sn bronzes<br>Delta phase<br>For Cu-Ga alloys, use values in parentheses. | 30 (100) mL<br>10 (8) mL<br>10 (25) mL | distilled water<br>hydrochloric acid 32%<br>nitric acid 65% | Seconds to minutes<br>!!! See section 4.4 in the Appendix. |
| Cu m13 | Cu alloys<br>Cu-Si alloys | 30 mL<br>1 g<br>50 mL<br>20 mL | distilled water<br>potassium hydroxide<br>ammonium hydroxide solution 25%<br>hydrogen peroxide 30% | 3–60 s<br>Follow mixing sequence as described! |
| Cu m14 | Cu-Sn-Ag dental alloys<br>Beta phase (Ag-Sn) is attacked. | 100 mL<br>3 mL<br>2 g<br>1 g | distilled water<br>sulfuric acid 95–97%<br>potassium dichromate<br>sodium fluoride<br>(Crowell's reagent) | Seconds to minutes<br>!!! See section 4.4 in the Appendix. |
| Cu m15 | Etches the (100) texture of Cu | 100 mL<br>20 g | distilled water<br>ammonium persulfate | Seconds to minutes<br>Use fresh only!<br>Heat if necessary. |
| Cu m16 | Cu and Cu alloys with soft solders<br>Solders and welds with brass<br>Epsilon phase<br>Segregations<br>Diffusion zones in solders<br>Alpha/beta brasses<br>Color etchant | 100 mL<br>5 g<br><br>**Klemm stock solution:**<br>300 mL<br>1000 mL | Klemm stock solution<br>potassium disulfite<br>(Klemm's reagent II)<br><br><br>distilled water (40 °C)<br>sodium thiosulfate | Seconds to minutes<br>Time will vary for different materials.<br>Sample surface must be wetted before etching (wet etching). |

(continued)

## Microetching (continued)

| No. | Material | Etchant | Conditions |
|---|---|---|---|
| Cu m17 | Pure Cu. Leaded bronze<br>Bronzes. Alpha/beta brasses<br>Al bronze<br>Be bronze<br>Color etchant | 100 mL distilled water<br>40 g potassium disulfite<br>11 mL Klemm stock solution (see Cu m16) (Klemm's reagent III) | 30 s to 5 min<br>Time will vary for different materials.<br>Sample surface must be wetted before etching.<br>Sample cannot be re-etched. |
| Cu m18 | Cu. Brasses<br>Bronzes | 60 mL ethanol 96%<br>30 mL distilled water<br>10 mL hydrochloric acid 32%<br>2 g iron (III) chloride | 1–2 min<br>Suitable for polarized light examination |

## Electrolytic Etching

| No. | Material | Etchant | Conditions |
|---|---|---|---|
| Cu Em1 | Al bronzes<br>Cu-Be alloys | 100 mL distilled water<br>1 g chromium (VI) oxide | 3–6 s<br>6 V DC<br>Al cathode<br>!!! See section 4.4 in the Appendix. |
| Cu Em2 | Beta brass<br>Alpaca<br>Bronzes<br>Monel metal<br>Cu-Ni alloys | 950 mL distilled water<br>50 mL sulfuric acid 95–97%<br>15 g iron (II) sulfate<br>2 g sodium hydroxide | Up to 15 s<br>8–10 V DC<br>Cu cathode |
| Cu Em3 | Cu alloys. Cartridge brass<br>Tombac. Muntz metal<br>Free-machining brass | 90 mL distilled water<br>10 mL phosphoric acid 85% | 5–10 s<br>1–8 V DC<br>Cu cathode |

## Physical Etching

Two oxide phases may be present in oxygen-containing Cu and Cu alloys: CuO and/or $Cu_2O$. Their color in brightfield illumination is very similar (gray to bluish). Darkfield illumination or polarized light can be used to distinguish and identify these phases; the $Cu_2O$ phase will become garnet red.

Fe

### *1.9.10 Iron, Steel, and Cast Iron*

## Preparation

The preparation of ferrous materials does not usually give rise to particular difficulties. However, the pronounced multiplicity of properties and microstructures in this category requires different standard methods of preparation.

Unalloyed iron and low-carbon steels, as well as some austenitic steels deform easily, but they also form deformation layers. Steels with bainitic and martensitic microstructures and the cast iron materials (C content of approximately 2 to 5%), on the other hand, are tough and brittle.

Hardened steels should not be heated during preparation (e.g., sectioning) since this might change the microstructure. In addition, in materials that have undergone a forming process, the direction of deformation should be taken into consideration. Fiber orientation, elongated grains, banded structures, and nonmetallic inclusions are better judged in longitudinal sections, while edge and surface variations, such as decarburized layers, hardened zones, etc., are best examined in cross sections.

I. Pure iron and low-carbon steels (sheet steel for deep-drawing applications, structural steels, case-hardened steels).

Steels with medium C content (steels for hardening and tempering).

Microalloy steels (e.g., grain-refined steels)

Manganese steels.

*Grinding:*

a. Wet grinding on SiC papers down to 1000 grit.
b. Wet grinding on SiC papers down to 4000 grit.

*Polishing:*

a. Grinding of I.a. Diamond polishing with 6 and 3 μm particle size on a cotton cloth. Final polishing on a tightly woven velvet cloth with 0.3 μm alumina slurry.
b. Grinding of I.b. Diamond polishing with 1 μm diamond particle size on a cotton cloth. Final polish of very soft steels or when preparing for color etching with $SiO_2$ suspension on a high-nap velvet cloth.
c. Electrolytic polish with Fe EP1 and Fe EP5.

II. Unalloyed tool steels, cold-working steels, hot-working steels, high-speed steels.

Martensitic steels.

Alloyed, austenitic, and austenitic-ferritic steels (e.g., CrNi, CrNiMo, CrNiSi, CrNiMoN steels).

Alloyed, ferritic, and ferritic-austenitic steels (e.g., Cr, CrAl steels, duplex steels).

Austenitic cast alloys.

*Grinding:*

Wet grinding on SiC papers down to 1000 grit.

*Polishing:*

a. Diamond polish with 6 and 3 μm particle size on a hard silk cloth.

Final polish with $SiO_2$ suspension on an acid-resistant synthetic cloth.

b. Lapping on a bakelite and metal composite disk with 6 μm diamond particle size using an alcohol-based lubricant. Diamond polish with 3 μm particle size on a silk cloth. Final polish with 0.3 μm alumina or $SiO_2$ suspension on an acid-resistant velvet cloth.

c. Electrolytic polish with Fe EP2.

III. Gray cast iron, tempered cast iron, nodular cast iron.

*Grinding:*

Wet grinding on SiC papers down to 2400 grit.

*Polishing:*

Diamond polish with 6 μm particle size on a perforated synthetic cloth, followed by 1 μm particle size on a hard synthetic fiber cloth. Final polish with a diluted $SiO_2$ suspension on a short-nap velvet cloth for 1 min.

## Electrochemical Polishing Recipes

*Electrolytic Polishing*

| | | |
|---|---|---|
| Fe<br>EP1 | Phosphoric acid 85% | 0.75–2.0 V<br>Fe cathode |
| Fe<br>EP2 | 200 mL methanol 99.8%<br>100 mL nitric acid 65% | Few s<br>40–50 V<br>Stainless steel cathode<br>Do not store!<br>!!! See section 4.4 in the Appendix. |
| Fe<br>EP3 | 400 mL distilled water<br>600 mL sulfuric acid 95–97% | 2–6 min<br>1.5–6 V DC<br>Ni cathode<br>Pure iron and stainless steels |
| Fe<br>EP4 | 830 mL distilled water<br>620 mL chromium (VI) oxide | 2–10 min<br>1.5–9 V DC<br>Ni cathode<br>High-alloy steels<br>!!! See section 4.4 in the Appendix. |
| Fe<br>EP5 | 120 mL distilled water<br>700 mL ethanol 96%<br>100 mL butyl cellosolve<br>78 mL perchloric acid 60% | 30–90 s<br>30–40 V<br>Stainless steel cathode!<br>!! See section 4.4 in the Appendix. |

## Macroetching

Preparation: Grinding on SiC papers down to 500 grit.

| No. | Material | Etchant | Conditions |
|---|---|---|---|
| Fe M1 | All unalloyed and low-alloy steels<br>Microalloy steels<br>Manganese steels<br>Welds | 100 mL distilled water<br>10 g ammonium persulfate | Wipe vigorously with cotton after etching. |
| Fe M2 | Unalloyed and low-alloy steels<br>Carburized and decarburized zones<br>Segregations | 90 mL ethanol 96%<br>10 mL nitric acid 65% | 1–5 min<br>Deep etch |
| Fe M3 | Flow lines in low carbon $N_2$ steels<br>Converter steels | 100 mL distilled water<br>120–180 mL hydrochloric acid 32%<br>45–90 g copper (II) chloride<br>(Fry's reagent) | 5–20 min<br>Heat treat before etching at 150–200 °C for 5–30 min.<br>After etching immerse into hydrochloric acid 32%, rinse in water and neutralize in a 32% ammonium hydroxide solution. |
| Fe M4 | Unalloyed steel<br>Free-cutting steel. Presence, distribution, and direction of sulfide inclusions | 100 mL distilled water<br>5 mL sulfuric acid 95–97%<br>(Sulfur print)<br>(Baumann print) | Saturate photographic paper with the mixture and press well onto the sample surface. Rinse after exposure of 3 min, fix (photographic fixer), wash, and dry. S-containing regions show brown discoloration. |
| Fe M5 | Unalloyed and alloy steels<br>Segregations<br>Primary structures<br>Banded structure | 500 mL distilled water<br>500 mL ethanol 96%<br>42 mL hydrochloric acid 32%<br>30 g iron (III) chloride<br>1 g copper (II) chloride<br>0.5 g tin (II) chloride<br>Add hydrochloric acid last.<br>(Oberhoffer's reagent) | Seconds to minutes<br>Surface must be polished.<br>Rinse in a 4:1 mixture of ethanol and hydrochloric acid.<br>Regions containing fewer alloying elements appear dark. |
| Fe M6 | Phosphorus segregations in low-carbon steels. Banded structure<br>Primary structures<br>Grain contrast | 100 mL distilled water<br>9 g copper (II) ammonium chloride<br>(Heyn's reagent) | 2–10 min<br>Wipe Cu deposit off with cotton under running water. |
| Fe M7 | High-alloy, corrosion-resistant steels | 25 mL distilled water<br>10 g copper (II) ammonium chloride<br>After dissolution add:<br>50 mL hydrochloric acid 32%<br>15 g iron (III) chloride<br>(Adler's reagent) | Rub sample surface vigorously with cotton. |
| Fe M8 | Austenitic and heat-resistant steels<br>FeCrNi-cast alloys | 50 mL hydrochloric acid 32%<br>25 mL saturated aqueous copper (II) sulfate solution<br>(Marble's reagent) | Seconds to minutes |

## Microetching

| No. | Material | Etchant | Conditions |
|---|---|---|---|
| Fe m1 | Most common etchant for pure iron, carbon steels, low-alloy steels, and gray cast iron | 100 mL ethanol 96%<br>1–10 mL nitric acid 65%<br>(Nital reagent) | Seconds to minutes |
| Fe m2 | Etch effect as for m1<br>Might reveal grain boundaries better than m1. | 100 mL ethanol 96%<br>2–4 g picric acid<br>(Picral reagent) | Seconds to minutes<br>!!! See section 4.4 in the Appendix. |
| Fe m3 | Etches segregations in unalloyed and low-alloy steels<br>Identifies cementite<br>Manganese. Cast iron<br>Color etch | 100 mL Klemm stock solution<br>2 g potassium disulfite<br>(Klemm's reagent I)<br>**Klemm stock solution:**<br>300 mL distilled water (40 °C)<br>1000 g sodium thiosulfate | Approximately 1–2 min<br>5 min for cast iron<br>Wet sample before etching. |
| Fe m4 | Flow lines in low-carbon $N_2$ steels<br>Converter steels | 30 mL distilled water<br>40 mL hydrochloric acid 32%<br>25 mL ethanol 96%<br>5 g copper (II) chloride<br>(Fry's reagent) | Seconds to minutes<br>Heat treat at 150–200 °C before etching. After etching immerse into hydrochloric acid 32%, rinse in water and neutralize with 25% ammonium hydroxide solution. |
| Fe m5 | Unalloyed and low-alloy steels<br>Martensitic and bainitic microstructures<br>Ferritic Cr steels<br>Manganese steels<br>Soldered joints with brass solder | 100 mL Beraha I stock solution<br>1 g potassium disulfite<br>(Beraha's reagent I)<br>**Beraha stock solution I:**<br>1000 mL distilled water<br>200 mL hydrochloric acid 32%<br>24 g ammonium hydrogen fluoride | 5–20 s<br>Wet sample before etching.<br>Mixture can be stored up to 1–2 h.<br>Store stock solution in plastic bottle.<br>!!! See section 4.4 in the Appendix. |
| Fe m6 | Grain boundaries in heat treated Cr alloy steels | 85 mL ethanol 96%<br>1–10 mL hydrochloric acid 32%<br>1–5 mL nitric acid 65% | Up to several min<br>!!! See section 4.4 in the Appendix. |
| Fe m7 | Grain boundaries in martensitic microstructures | 80 mL ethanol 96%<br>10 mL nitric acid 65%<br>10 mL hydrochloric acid 32%<br>1 g picric acid<br>(If necessary add a wetting agent) | Seconds to minutes<br>!!! See section 4.4 in the Appendix. |
| Fe m8 | Austenite grain boundaries in case-hardened steels and steels for hardening and tempering<br>Segregations are heavily attacked. | 100 mL saturated aqueous picric acid 65% (possibly diluted)<br>80 mg copper (II) chloride<br>3 mL wetting agent | 30–60 s<br>At 75–85 °C<br>Remove the black deposit on the sample surface with a 25% ammonium hydroxide solution.<br>!!! See section 4.4 in the Appendix. |
| Fe m9 | Austenite grain boundaries in cold-worked steels | Same as for Fe m8 with 1 mL hydrochloric acid 32% | Seconds to minutes<br>!!! See section 4.4 in the Appendix. |
| Fe m10 | Austenite grain boundaries in case-hardened steels. | 25 mL distilled water<br>75 mL ethanol 96%<br>20 mL saturated alcohol picric acid solution<br>12 mL wetting agent<br>10 g iron (II) chloride<br>3 g tin (II) chloride | Seconds to minutes<br>!!! See section 4.4 in the Appendix. |

(continued)

## Microetching (continued)

| No. | Material | Etchant | Conditions |
|---|---|---|---|
| Fe m11 | Austenite grain boundaries in valve steels<br>Hardening and tempering steels, case-hardening steels | 50 mL ethanol 96%<br>10 mL wetting agent<br>1 mL ammonium hydroxide solution 25%<br>1 mL hydrochloric acid 32%<br>3 g picric acid<br>1 g copper (II) ammonium chloride | Seconds to minutes<br>!!! See section 4.4 in the Appendix. |
| Fe m12 | Austenite grain boundaries in cold-worked steels | 48 mL ethanol 96%<br>10 mL wetting agent<br>6 g picric acid<br>1 g copper (II) ammonium chloride | Seconds to minutes<br>!!! See section 4.4 in the Appendix. |
| Fe m13 | High-alloy Cr and CrNi steels<br>Austenitic castings. Sigma phase and ferrite | 100 mL distilled water<br>100 mL hydrochloric acid 32%<br>10 mL nitric acid 65%<br>0.3 mL Vogel's special reagent<br>(Stainless steel, V2A reagent) | Seconds to minutes<br>RT to 70 °C |
| Fe m14 | Austenitic CrNi steels<br>Austenite is colored. Delta ferrite remains white.<br>Color etchant | 100 mL distilled water<br>20 mL ammonium hydrogen fluoride<br>0.5 g potassium disulfite<br>(Lichtenegger's + Bloech's reagent) | Seconds to minutes<br>Wet sample before etching.<br>!!! See section 4.4 in the Appendix. |
| Fe m15 | Ferritic Cr steels<br>Color etchant | 100 mL distilled water<br>10 mL ammonium hydroxide solution 25%<br>1 g potassium disulfite<br>(Modified Lichtenegger's + Bloech's reagent) | Seconds to minutes<br>Wet sample before etching. |
| Fe m16 | Austenitic CrNi steels<br>Delta ferrite colors first and then the austenite, sigma phase remains white.<br>Color etchant | 100 mL Beraha II stock solution<br>1 g potassium disulfite<br>(Beraha's reagent II)<br>**Beraha II stock solution:**<br>800 mL distilled water<br>400 mL hydrochloric acid 32%<br>48 g ammonium hydrogen fluoride | 10–20 s<br>Wet sample before etching. Can be stored for 1–2 h. Store mixture in a plastic bottle.<br>!!! See section 4.4 in the Appendix. |
| Fe m17 | Ferritic Cr steels<br>Austenitic CrNi steels<br>Austenitic cast alloys | 45 mL glycerol 87%<br>15 mL nitric acid 65%<br>30 mL hydrochloric acid 32%<br>(Villela's reagent) | Seconds to minutes<br>Use fresh only!<br>!!! See section 4.4 in the Appendix. |
| Fe m18 | Alloyed Cr steels<br>Cast iron. $Fe_3C$ colors dark faster in alloys with more than 10 Cr than in those with lower Cr content. $(Fe,Cr)_7C_3$, $(FeCr)_{23}C_6$, and the iron phosphide phase are colored. | 100 mL distilled water<br>10 g potassium or sodium hydroxide<br>10 g potassium ferricyanide<br>(Murakami's reagent) | 2–20 min<br>20–50 °C<br>Use fresh only! |
| Fe m19 | High-alloy steels and castings.<br>Sigma phase is colored, austenite is not colored, ferrite turns yellow-brown.<br>Color etchant | 60 mL distilled water<br>30 g potassium hydroxide<br>30 g potassium ferricyanide | 20–40 s<br>Heat if necessary.<br>Use fresh only! |

(continued)

## Microetching (continued)

| No. | Material | Etchant | Conditions |
|---|---|---|---|
| Fe m20 | As for Fe m19<br>Color etchant | 100 mL distilled water<br>4 g sodium hydroxide<br>4 g potassium permanganate<br>(Groesbeck's reagent) | 20 s to several min<br>50 °C |
| Fe m21 | Steels with high Si content | 20–40 mL glycerol 87%<br>10 mL nitric acid 65%<br>20 mL hydrofluoric acid 40% | Seconds to minutes<br>!!! See section 4.4 in the Appendix. |
| Fe m22 | Primary austenite grain boundaries in low- and medium-alloy steels<br>Martensitic and bainitic microstructures | 145 mL distilled water<br>16 g chromium (VI) oxide<br>80 g sodium hydroxide | 20–60 min at 120 °C<br>Preheat sample before etching.<br>!!! See section 4.4 in the Appendix. |
| Fe m23 | Martensitic, stainless steels<br>Martensite turns black. Ferrite different colors. Austenite is not attacked. | 33 mL distilled water<br>33 mL ethanol 96%<br>33 mL hydrochloric acid 32%<br>1.5 g copper (II) chloride<br>(Kalling's reagent 1) | Seconds to minutes |
| Fe m24 | Stainless steels<br>Ferrite is attacked quickly. Carbides are not attacked. Austenite is only slightly attacked. | 100 mL ethanol 96%<br>100 mL hydrochloric acid 32%<br>5 g copper (II) chloride<br>(Kalling's reagent 2) | Seconds to minutes |
| Fe m25 | Chi phase ($M_{18}C$ or $Fe_{36}Cr_{12}Mo_{10}$) | 100 mL distilled water<br>20 g potassium hydroxide<br>20 g potassium ferricyanide | Use boiling. Chi phase turns black and is found preferentially in the grain-boundary intersects. Sigma phase turns dark gray. Electrolytic polish before etching. |
| Fe m26 | Cementite ($Fe_3C$) colors dark in alloys up to 10 Cr, but remains unattacked for higher Cr-containing steels. $(Fe,Cr)_7C_3$, $(FeCr)_{23}C_6$, WC, and VC are not colored. | 75 mL distilled water<br>25 g sodium hydroxide<br>2 g picric acid | 3–15 min at 50 °C<br>!!! See section 4.4 in the Appendix. |
| Fe m27 | Presence of $Fe_3P$ | 100 mL distilled water<br>10 g potassium hydroxide<br>10 g potassium ferricyanide<br>1 g tartaric acid | 3 min at 50 °C<br>Cementite will color later.<br>Use fresh only! |

## Electrolytic Etching

| No. | Material | Etchant | Conditions |
|---|---|---|---|
| Fe Em1 | Microstructures of Cr and CrNi steels. Sigma phase is dissolved, carbides are only slightly attacked. | 100 mL distilled water<br>10 g oxalic acid | 5–20 s<br>1.5–3 V DC<br>!!! See section 4.4 in the Appendix. |
| Fe Em2 | High-alloy steels<br>Only carbides are etched and not the sigma phase. | Ammonium hydroxide solution 32% | 30–60 s<br>1.5–6 V DC<br>Pt cathode |

(continued)

## Electrolytic Etching (continued)

| No. | Material | Etchant | Conditions |
|---|---|---|---|
| Fe Em3 | Cr and CrNi steels<br>Cementite is rapidly attacked, austenite less, and ferrite and phosphide least.<br>Carbide etch. Cast iron and alloy cast iron | 100 mL distilled water<br>10 g chromium (VI) oxide | 3–60 s<br>3–6 V DC<br>Pt cathode<br>!!! See section 4.4 in the Appendix. |
| Fe Em4 | Presence of sigma phase<br>Sigma colors first, yellow to dark brown; ferrite colors later; longer etching also develops carbides. | 100 mL distilled water<br>40 g sodium hydroxide | 5–60 s<br>1–3 V DC<br>Pt cathode |
| Fe Em5 | Cr steels | 100 mL distilled water<br>10 mL Murakami's reagent (see Fe m18) | 6 s<br>2 V DC<br>Stainless steel cathode |

## Iron-Oxide Layers on Iron

| No. | Material | Etchant | Conditions |
|---|---|---|---|
| OFe m1 | FeO | 10 mL thioglycolic acid 5%<br>5 mL aqueous dipotassium phthalate solution 5%<br>2 mL aqueous ammonium citrate solution<br>3 mL aqueous citric acid solution 5% | 30–60 s<br>Swabbing |
| OFe m2 | FeO (multicolored/blue)<br>$Fe_2O_3$ (white)<br>$Fe_3O_4$ (beige) | 14 mL distilled water<br>85 mL ethanol 96%<br>1 mL bromine thymol blue | 2 h |
| OFe m3 | FeO (dull green-gray)<br>$Fe_2O_3$ (white)<br>$Fe_3O_4$ (beige) | 9 mL distilled water<br>90 mL ethanol 96%<br>1 mL bromine phenol red | 2 h |
| OFe m4 | FeO (brown-red/orange-brown)<br>$Fe_3O_4$ (beige blue/ light green) | 100 mL ethanol 96%<br>4 mL hydrochloric acid 32%<br>0.5 mL selenic acid 96% (Beraha's reagent) | 30–40 s<br>Mix chemicals in the prescribed sequence!<br>!!! See section 4.4 in the Appendix. |
| OFe m5 | $Fe_2O_3$<br>$Fe_3O_4$ and Fe are not attacked. | 10 mL distilled water<br>5 mL aqueous nitric acid solution 1%<br>5 mL aqueous citric acid solution 5%<br>5 mL aqueous thioglycolic acid solution 5% | 15–60 s<br>Swabbing. |
| OFe m6 | $Fe_2O_3$<br>$Fe_3O_4$ is not attacked. | 5 mL aqueous citric acid solution 10%<br>5 mL aqueous sodium thiocyanate solution 10% | 45–90 s |
| OFe m7 | $Fe_3O_4$ | (a) 15 mL distilled water<br>5 mL formic acid (98–100%)<br>(b) 15 mL distilled water<br>5 mL fluoboric acid 35% | 5 s<br>Swabbing<br>Then 2 s with (b)<br>!!! See section 4.4 in the Appendix. |

## Electrolytic Etching of Iron Oxide Layers

| No. | Material | Etchant | Conditions |
|---|---|---|---|
| OFe Em1 | $Fe_3O_4$<br>Fe and $Fe_2O_3$ are not attacked. | 10 mL aqueous thioglycolic acid solution 5%<br>5 mL aqueous dipotassium phthalate solution 5%<br>2 mL aqueous ammonium nitrate solution 5%<br>50 mL aqueous sodium chromate solution 0.5% | 15 s<br>Add sodium chromate solution just before use!<br>9 V DC, 2–4 mA/cm$^2$<br>Stainless steel cathode<br>!!! See section 4.4 in the Appendix. |

## Physical Etching

The different oxide layers that were formed by corrosion can be identified using polarized light (the individual layers are listed in sequence of their growth formation from the substrate):

FeO: (wüstite)
$Fe_3O_4$: (magnetite) appears dark
$Fe_2O_3$: (hematite) appears red
Wüstite is not stable below 500 °C and decomposes into magnetite and alpha Fe.
In Fe-Cr-Si steels, for example, a vapor deposition of ZnSe is recommended to improve the contrast of the individual phases.

## Removal of Rust Layers

| No. | Material | Etchant | Conditions |
|---|---|---|---|
| Fe R1 | Rust FeO(OH) removal from steels. (St37; 10CrMo9 10) | 100 mL water<br>15 g citric acid | 5–15 min |
| | | or: | |
| | | 100 mL water<br>15 g ascorbic acid | 5 min at RT |
| | | Both solutions can be used in the ultrasonic cleaner. Better results at shorter times are obtained by increasing the solution temperature and rinsing several times during the treatment. | |
| Fe R2 | Removal of cinder layers from steels. (10CrMo9 10; XCrMoV12 1; Incoloy 800H) | 50 mL water<br>10 mL Tickopur RW77 (alcohol concentration with ammonium hydroxide) | Up to several hours at RT<br>In the ultrasonic bath<br>Intermediate rinsing with alcohol |
| | | or: | |
| | | 160 mL distilled water<br>40 mL hydrochloric acid 32%<br>0.4 g hexamethylene tetramine | 1–15 min at RT<br>In the ultrasonic bath. Better results at higher temperature and with several intermediate rinsing steps |

### 1.9.11 *Semiconductors (Germanium, Silicon, Selenium, Tellurium, $A_{III}B_{V}$, and $A_{II}B_{VI}$ Compounds)*

**Preparation**

The objective for the preparation of semiconductors is often to produce crystallographically oriented sections from single crystals without any remaining surface deformation and also to reveal structural defects, e.g., dislocation density or dislocation distribution, and to identify differences of doping (growth streaks and seed cells). Samples that are too small for easy handling should be mounted before sectioning. The sectioning of semiconductors according to their crystal orientation is an elaborate procedure that is not discussed in this book.

The samples that were sectioned from the wafers are glued onto a sample holder. For careful removal of the deformed surface, the sample is not ground, but lapped with fine-grained abrasives. This can be done either manually or with automatic equipment. Plane glass disks are used for manual lapping and rotating disks of cast iron, brass, aluminum, and Pb-Sn disks are used for automatic lapping. Lapping abrasives are emery, corundum, and silicon carbide. Suitable polishing compounds are aluminum oxide, chromium oxide, cerium oxide, and magnesium oxide with a particle size of <1 μm, a short-nap and a dense velvet polishing cloth is used. Contrary to lapping, lower pressures at higher speed with small amounts of polishing suspension are recommended for manual and automatic polishing.

I. Ge pure 5N/6N, doped with Al, P, In.

*Lapping:*

On a glass plate with emery of 15 and 5 μm particle size and water.

*Polishing:*

a. On a Pb-Sn lapping disk with 3 μm diamond particle size, then on typing paper with 3 and 1 μm diamond (using any flat substrate), followed by a brief polish on a short-nap velvet cloth with 0.25 μm diamond particle size. Final polish on a chemical-resistant cloth with $SiO_2$ suspension and a 10% aqueous sodium perchlorate solution at a 1:1 mixture, or with diluted $SiO_2$ suspension and a few drops of bromine-methanol solution (1 mL bromine to 10 mL methanol).
b. For polycrystalline material:
1 μm diamond particle size on a Pb-Sn disk, 1 μm diamond particle size on a short-nap velvet cloth, then coat a chemically resistant cloth with paraffin and with $SiO_2$ suspension and 6 drops of 10% aqueous sodium perchlorate solution; polish with pressure.
c. Chemical polish with Ge CP1.

II. Si pure 5N/6N, doped with B.

*Grinding:*

Wet grinding with SiC papers down to 1000 grit; then with 3 μm diamond particle size on a woven steel cloth and an alcohol-based lubricant.

or:

*Lapping:*

With fine grit, 4 μm emery, and water on a glass plate.

*Polishing:*

a. $SiO_2$ suspension on a perforated synthetic cloth, then on a chemically resistant synthetic cloth.
b. With 3 and 1 μm diamond particle size on a Pb-Sn lapping disk, then with 1 μm diamond particle size on paper. Final polish with $SiO_2$ suspension on a chemically resistant synthetic cloth.
c. Electrolytic polish with Ge EP1.
d. Chemical polish with Ge CP2.

III. Se pure and doped with Br and Cl, alloyed with 0.01 to 0.1 As and Bi for rectifiers. Very toxic (!!! See section 4.4 in the Appendix.).

Te pure, for glasses and glazes, very toxic (!!! See section 4.4 in the Appendix.).

*Lapping:*

With 15 and 4 μm emery particle size and water on a glass plate.

*Polishing:*

With 0.3 μm alumina slurry on a hard silk cloth.

IV. $A_{III}B_{VI}$ compounds, e.g., GaP, InP, GaAs.

IV.1 GaP and InP must be prepared under a fumehood because of possible generation of phosphines.

*Lapping:*

With 4 μm emery particle size and ethylene glycol on a glass plate.

*Polishing:*

a. With 3 μm diamond particle size on paper placed onto a glass plate; then on optical paper with 1 μm diamond particle size; then with $SiO_2$ suspension mixed at a 1:1 ratio with a 10% aqueous sodium perchlorate solution on a chemically resistant synthetic cloth.

b. After lapping polish on a chemically resistant synthetic cloth first with a mixture of 2 parts of 0.3 μm alumina and 1 part of 10% aqueous sodium perchlorate solution, then with $SiO_2$ suspension and 10% sodium perchlorate solution at a 1:1 ratio.

IV.2 GaAs crystals with {110} and {111} crystal orientation or cleavage planes are very difficult to polish. GaAs is very toxic (!!! See section 4.4 in the Appendix.).

*Grinding:*

Wet grinding on SiC papers down to 1000 grit; then fine grinding with 3 μm diamond particle size and an alcohol-based lubricant on a woven steel cloth.

*Polishing:*

a. With a 10% aqueous sodium perchlorate solution on a chemically resistant synthetic cloth.

b. A 3 μm diamond polish on a hard silk cloth is recommended for {111} crystal orientations; then with a 10% aqueous sodium perchlorate solution on chemically resistant synthetic cloth.

V. $A_{II}B_{VI}$ compounds, e.g., CdTe, HgTe.

*Lapping:*

With 4 μm grit emery and water on a glass plate.

*Polishing:*

Polish with 3 and 1 μm diamond particle size on paper placed on a glass plate. Final polish with a compound used for polishing plastic compounds (see section 3.1 and Table 3.4) and ethylene glycol on a short-nap velvet cloth and with a 10% aqueous sodium perchlorate solution.

VI. Semiconductors are used for microchips, conductive boards, and printed circuit boards.

The semiconductors with ceramics and metal bonds are difficult to prepare because of the presence of phases with different hardness and brittleness. These bonded materials are examined metallographically for defects such as cracks, pores, and gaps, and the adhesion between the individual layers. Taper sections are recommended for very thin layers; this technique is particularly used for thin-film components.

VI.1. Printed circuits and printed circuit boards.

*Sectioning:*

Individual samples can easily be sectioned with a SiC cutoff wheel. If many samples are to be sectioned, it is faster to punch the individual samples out of the circuit boards.

*Mounting:*

For automatic preparation, the samples should be mounted in a device with reference holes to maintain the same level in all samples.

*Grinding:*

Wet grinding with SiC papers from 800/1000 grit down to 1200 grit.

*Polishing:*

a. With 6 μm diamond particle size on a hard synthetic cloth, then with 3 μm diamond particle size on a hard silk cloth, then with 1 μm diamond particle size on a short-nap velvet cloth. Final polish, if necessary, with $SiO_2$ suspension. Use short times and low pressures.
b. Etch between polishing steps with 40 mL hydrogen peroxide 30% and 60 mL ammonium hydroxide solution 25% (!!! See section 4.4 in the Appendix.).

VI.2 Transistors and resistors (ceramics with metal coatings).

*Sectioning:*

Care must be taken when clamping the samples for sectioning with the cutoff wheel. Use elastic materials for spacers. The adhesion between the ceramic and metal layers is usually low, and the layers can easily peel off the ceramic material.

*Mounting:*

Use epoxy resins for ceramics with lower hardness, add hard material for balance with harder ceramics.

*Grinding:*

a. Wet grinding with SiC papers of finer grit when the ceramic portion of the sample is small.
b. If the ceramic portion of the sample is large, use a Bakelite bonded diamond abrasive disk.

*Lapping:*

With 6 μm diamond particle size on a medium-hard lapping disk.

*Polishing:*

a. With 3 μm diamond particle size on a hard synthetic cloth and with 1 μm diamond particle size on a hard silk cloth. Final polishing on a short-nap velvet cloth with $SiO_2$ suspension. This will polish mainly the metal coatings. Use short polishing times.
b. For good distinction between solder and metallized coatings, the sample is chemically polished for a short time with 0.5 μm alumina and intermittently etch polished with a 10% aqueous ammonium tartrate solution on a chemically resistant synthetic cloth.

## Electrochemical Polishing Recipes

*Electrolytic Polishing*

| | | |
|---|---|---|
| Ge EP1 | 590 mL methanol 99.8%<br>350 mL 2-butoxy ethanol 99%<br>60 mL perchloric acid 60% | 30–60 s<br>25–35 V DC<br>Stainless steel cathode<br>!!! See section 4.4 in the Appendix. |

*Chemical Polishing*

| | | |
|---|---|---|
| Ge CP1 | 25 mL nitric acid 65%<br>15 mL hydrofluoric acid 40%<br>15 mL glacial acetic acid<br>3–4 drops bromine | 5–10 s<br>!!! See section 4.4 in the Appendix. |
| Ge CP2 | 20 mL nitric acid 65%<br>5 mL hydrofluoric acid 40% | 5–10 s<br>!!! See section 4.4 in the Appendix. |

## Macroetching

| No. | Material | Etchant | Conditions |
|---|---|---|---|
| Ge M1 | Si-B-P alloys<br>Grain-boundary etchant | 100 mL distilled water<br>5 g sodium hydroxide | Seconds to minutes<br>Use hot. |
| Ge M2 | Pure Ge and Ge alloys | 50 mL distilled water<br>50 mL hydrochloric acid 32%<br>20 g iron (III) chloride | 60 s<br>Use boiling. |

## Microetching

| No. | Material | Etchant | Conditions |
|---|---|---|---|
| Ge m1 | Pure Si and Ge and their alloys<br>InSb | 5 (90) mL distilled water<br>25 (5) mL nitric acid 65%<br>25 (5) mL hydrofluoric acid 40% | 5–20 s<br>Use values in parentheses for Si.<br>!!! See section 4.4 in the Appendix. |
| Ge m2 | Ge-In alloys with Ag, Au, Bi, and Cu additions<br>Grain-boundary etchant | 100 mL ethanol 96%<br>5 mL hydrochloric acid 32%<br>1 g picric acid | Seconds to minutes<br>!!! See section 4.4 in the Appendix. |

(continued)

## Microetching (continued)

| No. | Material | Etchant | Conditions |
|---|---|---|---|
| Ge m3 | Pure Si and Ge and their alloys<br>InSb<br>Etch figures of the (111) plane.<br>Etches *p-n* transitions | 10 mL hydrofluoric acid 40%<br>10 mL nitric acid 65% | Seconds to minutes<br>!!! See section 4.4 in the Appendix. |
| Ge m4 | Pure Si<br>Pure Te<br>Pure Se | 100 mL distilled water<br>50–100 g sodium hydroxide | 2–20 min<br>Use hot for Te if necessary. |
| Ge m5 | Se, Ge, and their alloys<br>GaSb and InSb<br>Dislocations on the (100) and (111) crystal planes of Si<br>InAs | 45.5 mL nitric acid 65%<br>27 mL hydrofluoric acid 48%<br>27 mL glacial acetic acid<br>0.5 mL bromine | 3–25 s<br>Mix 30 min before use; dissolve the bromine first in the glacial acetic acid.<br>!!! See section 4.4 in the Appendix. |
| Ge m6 | Ge and Ge alloys<br>Grain-contrast etchant<br>Can also be used for macroetching | 50 mL distilled water<br>50 mL hydrochloric acid 32%<br>20 g iron (III) chloride | Seconds to minutes<br>Use boiling. |
| Ge m7 | Ge and Ge alloys<br>GaAs, InAs, and AlAs<br>Dislocations on the (111) planes<br>Grain-boundary etchant | 40 mL distilled water<br>40 mL hydrofluoric acid 40%<br>20 mL nitric acid 65%<br>2 g silver nitrate | 30 s to 2 min<br>!!! See section 4.4 in the Appendix. |
| Ge m8 | Pure Ge, Te, and Se<br>Tellurides, selenides, and Zr silicides | nitric acid 65% | Seconds to minutes<br>If necessary, dilute with distilled water and/or add hydrochloric acid. |
| Ge m9 | Se, Ge, and their alloys<br>InAs, InSb, InP, AlSb, GaAs, GaSb, ZnTe, and CdTe | 40 mL distilled water<br>10 mL hydrofluoric acid 40%<br>10 mL hydrogen peroxide 30% | 1–3 min<br>!!! See section 4.4 in the Appendix. |
| Ge m10 | SbTe and BiTe<br>Detection of Te<br>InP | 100 mL distilled water<br>10 g potassium or sodium hydroxide<br>10 g potassium ferricyanide (Murakami's reagent) | Seconds to minutes<br>Use fresh only! |
| Ge m11 | Presence of $Si_2O$ in Si | 90 mL distilled water<br>15 mL hydrochloric acid 32%<br>10 mL hydrofluoric acid 40% | Seconds to minutes<br>!!! See section 4.4 in the Appendix. |
| Ge m12 | GaAs and GaP<br>Etches A and B planes in GaAs | 20 mL sulfuric acid 95–97%<br>20 mL hydrofluoric acid 40%<br>80 mL hydrogen peroxide 30% | 5 min at 90–100 °C<br>A planes: layers of Ga atoms show etch pits.<br>B planes: layers of As atoms show fewer etch pits.<br>!!! See section 4.4 in the Appendix. |
| Ge m13 | BiTe and GaP | 50 mL distilled water<br>25 (5) mL nitric acid 65%<br>25 (5) mL hydrochloric acid 32% | Seconds to minutes<br>Possibly without water<br>Use values in parentheses for BiTe. |

## Electrolytic Etchants

| No. | Material | Etchant | Conditions |
|---|---|---|---|
| Ge Em1 | Pure Ge and Ge alloys<br>Grain-boundary etchant | 1000 mL distilled water<br>100 g oxalic acid | 10–20 s<br>4–6 V DC<br>Stainless steel cathode<br>!!! See section 4.4 in the Appendix. |
| Ge Em2 | BiTe and BiSe | 570 mL distilled water<br>56 g sodium hydroxide<br>48 g tartaric acid | Seconds to minutes<br>0.5 A/cm$^2$. Stainless steel cathode |

## *1.9.12 Hafnium and Zirconium*

| Hf |
|---|
| Zr |

**Preparation**

I. Pure zirconium (99.99) is used as structural material for the chemical industry and for fuel tubes in light and heavy water reactors. Pure hafnium is used for control rods in nuclear reactors.

High-purity Zr and Hf are soft, but the hardness increases rapidly with oxygen contents above 800 ppm. At room temperature, both metals exhibit the strongly polarizing alpha phase. The materials deform very easily and may generate deformation twins.

*Grinding:*

a. Ultramilling with a tungsten carbide knife (diamond knives cause surface roughness).
b. Wet grinding on new SiC papers down to 1200 grit. If necessary coat the finer grits with wax. Longer polishing times are required to remove the smeared layer. Fine grinding with SiC papers down to 4000 grit is recommended.

*Polishing:*

a. With diamond or alumina slurry to the finest particle size, using a rapidly rotating wheel with long polishing times on a medium soft cloth. If necessary, etch between polishing steps with Hf m1, Hf m2, or Hf m6.
b. Chemical-mechanical polish after fine grinding with 4000 grit or after diamond polishing down to 1 µm. Because the etchants contain hydrofluoric acid, an enclosed polishing dish should be used and safety equipment for eyes and hands are mandatory!!!! A vibratory polisher is recommended.
   The solution used for Zr: add 10 drops of hydrofluoric acid 40% and 5 drops of nitric acid 65% to 100 mL alumina oxide suspension, then polish for 60 to 180 min. For Hf, use the following solution: add 4 to 30 drops of hydrofluoric acid 40% to 100 mL alumina suspension and polish on a cotton cloth.
c. Electrolytic polish with Hf EP1, Hf EP2, and Hf EP3 for Zr and Hf EP2 for Hf.
d. Chemical polish with Hf CP1 or Hf CP2.

II. Zircaloy 2, Zircaloy 4 (Zr with Sn, $O_2$ and Fe, Cr, Ni). Zr-2.5Nb. Zr-4.5Hf. Used in the chemical industry for containers and water pipes, and water steam pipes used at high temperatures and high pressures.

Additional alloying elements such as Hf, Sn, $N_2$, and $O_2$ form solid solutions in the alpha phase and stabilize it; Fe, Cr, and Ni are only soluble above 852 °C in the beta phase and precipitate as intermetallic compounds in the RT stable alpha microstructure.

*Grinding:*

As I.b.

*Polishing:*

a. As I.a.
b. As I.b.
c. Etch polish Hf-Cr and Hf-V with a 5% aqueous chromium (VI) oxide solution and gamma alumina suspension (!!! See section 4.4 in the Appendix.).

## Electrochemical Polishing Recipes

*Electrolytic Polishing*

| | | |
|---|---|---|
| Hf EP1 | 90 mL glacial acetic acid<br>10 mL perchloric acid 60% | 30 s to 2 min<br>12–18 V DC<br>Stainless steel cathode<br>!!! See section 4.4 in the Appendix. |
| Hf EP2 | 96 mL distilled water<br>5 mL hydrofluoric acid 40% | For pure Zr<br>80–120 s. 5 V DC<br>Stainless steel cathode<br>!!! See section 4.4 in the Appendix. |
| Hf EP3 | 100 mL distilled water<br>6 mL nitric acid 65%<br>6 mL hydrofluoric acid 40% | For pure Zr<br>60–80 s. 5 V DC<br>Stainless steel cathode<br>!!! See section 4.4 in the Appendix. |

*Chemical Polishing*

| | | |
|---|---|---|
| Hf CP1 | 45 mL distilled water or hydrogen peroxide 30%<br>45 mL nitric acid 65%<br>10 mL hydrofluoric acid 40% | Several times 5–10 s<br>!!! See section 4.4 in the Appendix. |
| Hf CP2 | 45 mL lactic acid 90%<br>45 mL nitric acid 65%<br>8 mL hydrofluoric acid 40% | Seconds to minutes<br>!!! See section 4.4 in the Appendix. |

## Macroetching

| No. | Material | Etchant | Conditions |
|---|---|---|---|
| Hf M1 | Pure Hf<br>Zircaloy 2 | 70 mL distilled water<br>30 mL nitric acid 65%<br>5 mL hydrofluoric acid 40% | Seconds to minutes<br>Swabbing<br>!!! See section 4.4 in the Appendix. |
| Hf M2 | Pure Zr, pure Hf, and their alloys with high additions | 45 mL distilled water<br>45 mL nitric acid 65%<br>10 mL hydrofluoric acid 40% | Seconds to minutes<br>Swabbing<br>For low-alloy elements, use ethanol instead of water.<br>!!! See section 4.4 in the Appendix. |
| Hf M3 | Pure Zr, pure Hf, and their alloys with low additions | 45 mL hydrogen peroxide 30%<br>45 mL nitric acid 65%<br>10 mL hydrofluoric acid 40% | Seconds to minutes<br>Swabbing<br>!!! See section 4.4 in the Appendix. |

## Microetching

| No. | Material | Etchant | Conditions |
|---|---|---|---|
| Hf m1 | Pure Hf. Pure Zr<br>Low-alloy Hf and Zr<br>Zr-U, Zr-Al, and Hf-Re alloys | 45 mL distilled water<br>45 mL nitric acid 65%<br>10 mL hydrofluoric acid 40% | 5–20 min. Suitable for polarized light examination<br>!!! See section 4.4 in the Appendix. |
| Hf m2 | Hf- and Zr-base alloys<br>Zr-Nb alloys | 30 mL hydrochloric acid 32%<br>30 mL hydrofluoric acid 40%<br>15 mL nitric acid 65% | 3–10 s<br>Swabbing, 2 min immersion etching<br>!!! See section 4.4 in the Appendix. |
| Hf m3 | Zr-Be and Zr-H alloys<br>Pure Zr. Zircaloy | 100 mL distilled water<br>10 mL hydrofluoric acid 40%<br>(Concentration variable) | Seconds to minutes<br>!!! See section 4.4 in the Appendix. |
| Hf m4 | Hf-W alloys | 98 mL nitric acid 65%<br>2 mL hydrofluoric acid 40% | 2–40 s<br>!!! See section 4.4 in the Appendix. |
| Hf m5 | Pure Zr<br>Zr-U alloys | 30 mL lactic acid 90%<br>30 mL nitric acid 65%<br>2 mL hydrofluoric acid 40% | Seconds to minutes. Use 6–10 drops of hydrofluoric acid for Zr-U alloys.<br>Solution decomposes; do not store!<br>!!! See section 4.4 in the Appendix. |
| Hf m6 | Hf-base alloys<br>For Zr-base alloys, use concentration in parentheses. | 200 (100) mL distilled water<br>5 mL hydrofluoric acid 40%<br>2 mL aqueous silver nitrate solution 5% | 5–60 s<br>!!! See section 4.4 in the Appendix. |
| Hf m7 | Hydride in Zr-Nb alloys | 45 mL lactic acid 90%<br>45 mL nitric acid 65%<br>8 mL hydrofluoric acid 40% | 10–20 s<br>Swabbing<br>!!! See section 4.4 in the Appendix. |
| Hf m8 | Zr and high-alloy Zr | 45 mL nitric acid 65%<br>45 mL hydrogen peroxide 30% | 5–10 s<br>!!! See section 4.4 in the Appendix. |
| Hf m9 | Zr-Cu, Zr-Nb, Zr-Ni, Zr-Si, Zr-Sn, Zr-Th alloys with Al, Be, Fe, Ni, Si | 90 mL nitric acid 65%<br>10 mL hydrofluoric acid 40%<br>(Concentration variable) | Seconds to minutes<br>!!! See section 4.4 in the Appendix. |
| Hf m10 | Zr-B, Zr-Fe, Zr-Ni, Zr-Mo, Zr-Sn, Zr-U alloys | 85 (45) mL glycerol 87%<br>10 (45) mL nitric acid 65%<br>5 (10) mL hydrofluoric acid 40 %<br>(Concentration variable) | Seconds to minutes<br>Use concentration in parentheses for alloys with low additions.<br>Do not store!<br>!!! See section 4.4 in the Appendix.<br>! Explosive! |

## Electrolytic Etching

| No. | Material | Etchant | Conditions |
|---|---|---|---|
| Hf Em1 | Pure Zr and Zr alloys | 100 mL glycerol 87%<br>10 mL nitric acid 65%<br>5 mL hydrofluoric acid 40% | 1–10 min<br>9–12 V DC<br>Pt cathode. Must be cooled! Do not store!<br>!!! See section 4.4 in the Appendix.<br>! Explosive! |

## Anodizing

The sample should be lightly etched before anodizing. A stainless steel beaker is the cathode.

| No. | Material | Etchant | Conditions |
|---|---|---|---|
| Hf Am1 | Pure Zr. Pure Hf<br>Grain contrast | 60 mL ethanol 96%<br>25 mL distilled water<br>20 mL glycerol 87%<br>10 mL lactic acid 90%<br>5 mL phosphoric acid 85%<br>2 g citric acid | 10–20 s<br>15–20 V DC. Anodize until golden color.<br>Suitable for polarized light. Do not store!<br>!!! See section 4.4 in the Appendix.<br>! Elemental hydrogen may be formed! |
| Hf Am2 | Zr alloys with intermetallic phases, oxides, nitrides, and hydrides | As above | 10–20 s<br>110–115 V DC<br>Anodize until purple red to blue (examination with brightfield).<br>!!! See section 4.4 in the Appendix. |
| Hf Am3 | Zircaloy with precipitates | As above | 10–20 s<br>180 V DC<br>Anodize until light red.<br>!!! See section 4.4 in the Appendix. |

Hg

### *1.9.13 Mercury (Amalgams)*

## Preparation

Mercury and amalgams are toxic (!!! See section 4.4 in the Appendix.).

I. Amalgams are Hg alloys produced by mixing Hg with a powder mixture of Ag, Sn, Cu, and Zn (26 to 69 Ag, 24 to 31 Sn, and 3 to 39 Cu; mixed with 50 to 70 Hg). Used for tooth fillings.

Amalgams are heat sensitive. They have a multiphase microstructure with hard and soft constituents. The preparation is difficult; e.g., deep scratches may persist, and the material tends to smear. Only light pressures should be used, and the samples should be well cooled during preparation. Microscopic examination should be carried out as soon as the sample preparation is completed.

*Mounting:*

Mount in cold-setting plastic materials.

*Grinding:*

a. Wet grinding on used SiC papers down to 1200 grit. Then continue using microcut papers.
b. Ultramilling may replace grinding and polishing.

*Polishing:*

a. Polish with 3 and 0.25 μm diamond particle size for approximately 5 to 15 min. Then with 0.05 μm alumina slurry or with magnesia slurry on a velvet or felt cloth with low pressure and at high speed. Rinse only with distilled water and dry with a stream of cold air.
b. Etch polishing with a microetchant.

II. Mg amalgams.

*Grinding:*

Grind on SiC papers down to 1200 grit. Avoid any contact with water or other moisture; Mg forms water-soluble phases.

*Polishing:*

For polishing, use very fine alumina slurries diluted with Vaseline. The samples are rinsed with trichloroethylene (!!! in a fume hood). Chemical etching is not necessary. Amalgams are coated with a thin layer of transparent lacquer or with transparent tape to prevent the corrosion of amalgams by the humidity in the air.

## Microetching

| No. | Material | Etchant | Conditions |
|---|---|---|---|
| Hg m1 | Hg-Sn and Hg-Sn-Cu alloys | Glacial acetic acid | 15 min |
| Hg m2 | Ag-Sn-Hg alloys | 100 mL ammonium hydroxide solution 32%<br>25 mL hydrogen peroxide 30%<br>(If necessary dilute with 100 mL distilled water) | Seconds to minutes |
| Hg m3 | For most amalgams | (a) 90 mL distilled water or ethanol 96%<br>10 mL hydrochloric acid 32%<br>(b) 90 mL distilled water<br>10 mL aqueous iron (III) chloride solution 5%<br>1 mL hydrochloric acid 32% | 1–3 min<br>Short immersion<br>(a) is usually sufficient.<br>Etching in (a) and (b) improves the contrast. |
| Hg m4 | Ag-Sn-Hg alloys (hardened) | 100 mL distilled water<br>30 mL nitric acid 65% | Seconds to minutes<br>Time depends on the Hg content. High Hg needs longer times. |

Etchants recommended for Hg alloys are also suitable for amalgams, but only when diluted.

**Physical Etching**

Reactive sputtering using a Au or Fe cathode with oxygen for 8 to 20 min. Interrupt after each minute for 1 min to avoid heating of the sample.

Mg

## *1.9.14 Magnesium*

**Preparation**

Magnesium dust and small chips are ignitable (!!! See section 4.4 in the Appendix.). Be careful when cutting, use a coolant.

I. Technically pure magnesium (99.9) is not used as an industrial material because of its lack of mechanical properties.

Pure Mg is difficult to prepare. It is very sensitive to deformation and because of its low hardness, grinding and polishing compounds are easily pressed into the surface.

*Mounting:*

Use only cold-setting materials.

*Grinding:*

a. Ultramilling with a diamond knife and an alcohol-based lubricant.
b. Wet grinding on SiC papers, starting with the finest paper down to 1200 grit. Use distilled water or a mixture of glycerol and ethanol 1:3. Increase grinding time with decreasing grain size. Use low speeds and low pressure. Rinse only with ethanol.

*Polishing:*

a. Rough polish with 5 μm alumina slurry in distilled water on a cotton cloth and a long polishing time. Final polishing with 0.05 μm alumina slurry on a short-nap velvet cloth treated with soap.
b. Electrolytic polish with Mg EP1.
c. Chemical polish with Mg CP1.

II. Mg-Al-Mn with or without Zn, Si.

Mg-Zn-Zr with or without Th.

Mg-Rare Earths-Zr with or without Zn, Ag.

Mg-Th-Zr with or without Zn.

Mg-Li-Al, Mg-Zn, Mg-Mn.

Application in the automobile and aviation industry, for instrumentation dashboards and formed sheets.

Mg alloys also require a careful sample preparation. Avoid using water, some phases will react with tap water!

*Grinding:*

On SiC papers down to 1200 grit, wet with distilled water or a 1:3 mixture of glycerol and ethanol, starting with the finest grit paper possible. Double the grinding times with decreasing grain size. Use low speed and low pressure. Rinse only with ethanol.

*Polishing:*

a. Rough polish with 5 μm alumina slurry in distilled water on a cotton cloth. Use long polishing times. Final polishing with 0.05 μm alumina slurry on a low-nap velvet cloth treated with soap.
b. Polish with 6, 3, and 1 μm diamond particle size on a hard synthetic cloth, a hard silk cloth, and a velvet cloth. Final polishing as in II.a.
c. After 3 μm diamond polish use a chemical-mechanical polish with a 0.05 μm alumina slurry and a solution of:
   20 mL aqueous calcium dichromate solution (2%)
   150 mL saturated aqueous boric acid solution
   15 drops of nitric acid 65%
d. Electrolytic polish with Mg EP2.
e. Chemical polish with Mg CP2. Samples must be ground down to 1200 grit.

## Electrochemical Polishing Recipes

*Electrolytic Polishing*

| | | | |
|---|---|---|---|
| Mg EP1 | 250 mL | distilled water | 2 min. 25–50 °C |
| | 300 mL | ethanol 96% | 10 V DC |
| | 400 mL | orthophosphoric acid 85% | Mg or Al cathode |
| | | | !!! See section 4.4 in the Appendix. |
| Mg EP2 | 100 mL | ethanol 96% | 20 °C |
| | 10 mL | butyl cellosolve | 23 V DC |
| | 20 g | sodium thiocyanate | Hg or Al cathode |
| | 10 g | sodium hydrogen phosphate of ethylene diamine tetra acetic (EDTA) | |

*Chemical Polishing*

| | | |
|---|---|---|
| Mg CP1 | Nitric acid 65% | Immersion at 3 s intervals<br>Polishing effect after 1 min<br>Then etching |
| Mg CP2 | 100 mL ethanol 96%<br>12 mL hydrochloric acid 32%<br>8 mL nitric acid 65% | 30 s<br>Grain boundaries are slightly etched.<br>!!! See section 4.4 in the Appendix. |

## Macroetching

| No. | Material | Etchant | Conditions |
|---|---|---|---|
| Mg M1 | Mg materials<br>Grain size, flow lines in cast and forged parts | 20 mL distilled water<br>50 mL 64% picric acid solution in ethanol 96%<br>20 mL glacial acetic acid.<br>(If necessary increase the amount of water) | 30 s to 3 min<br>Rinse off deposit with hot distilled water.<br>!!! See section 4.4 in the Appendix. |
| Mg M2 | Pure Mg<br>Internal defects in cast bars and ingots. Segregation<br>Flow lines in forged parts | 100 mL distilled water<br>10 mL glacial acetic acid | 30 s to 5 min<br>Swabbing |
| Mg M3 | Mg-Mn and Mg-Zr alloys<br>Internal defects in cast bars and ingots. Segregation and flow lines in forged parts | 100 mL distilled water<br>20 mL nitric acid 65% | 30 s to 5 min |

## Microetching

| No. | Material | Etchant | Conditions |
|---|---|---|---|
| Mg m1 | Pure Mg and most Mg alloys (as cast or forged) | 100 mL ethanol 96% or distilled water<br>1–8 mL nitric acid 65% (Nital reagent) | Seconds to minutes |
| Mg m2 | All Mg alloys in the as cast or heat treated condition<br>Grain-boundary etchant | 24 mL distilled water<br>75 mL ethylene glycol<br>1 mL nitric acid 65% | 30–60 s |
| Mg m3 | Pure Mg<br>Mg-Al, Mg-Al-Zn, Mg-Mn, Mg-Th-Zr, and Mg-Zn-Zr alloys. Also extrusion material<br>Grain-boundary etchant | 100 mL distilled water<br>2 g oxalic acid | 6–10 s<br>!!! See section 4.4 in the Appendix. |
| Mg m4 | Mg-Al, Mg-Mn and Mg-Mn Al-Zn alloys<br>Flow lines, grain size in castings | 90 mL distilled water<br>2–10 g tartaric acid | 10 s to 2 min<br>Dry polish only!<br>Swab vigorously with the etchant and rinse with ethanol, dry in a stream of cold air.<br>For Mg-Al alloys with >6 Al: add 20 g tartaric acid. |

(continued)

## Microetching (continued)

| No. | Material | Etchant | Conditions |
|---|---|---|---|
| Mg m5 | Technically pure Mg<br>Mg-Al-Zn, Mg-Zn-Th-Zr, and Mg-Rare Earth alloys<br>$Mg_{17}Al_{12}$ (dark) | 90 mL distilled water<br>10 mL hydrofluoric acid 40% (Hydrofluoric acid content may be lowered) | 3–30 s<br>Sample must be agitated.<br>!!! See section 4.4 in the Appendix. |
| Mg m6 | Mg-Al alloys<br>Grain-contrast etchant for heat treated castings<br>Flow lines in forgings | 85 mL distilled water<br>15 mL nitric acid 65%<br>12 g chromium (VI) oxide | 10–30 s<br>Increase water for alloys with high Al content.<br>!!! See section 4.4 in the Appendix. |
| Mg m7 | For most Mg materials and Mg alloys, as well as for castings and forgings | 19 mL distilled water<br>60 mL ethylene glycol<br>20 mL glacial acetic acid<br>1 mL nitric acid 65% | 1–30 s<br>Swabbing<br>Rinse sample with hot water. |
| Mg m8 | Pure Mg and Mg-Cu alloys<br>For Mg die-cast alloys | 100 mL ethanol 96%<br>2–11 g citric acid | Up to 30 s<br>Swabbing<br>Rinse sample with hot water. |
| Mg m9 | Mg-Si alloys<br>$Mg_2Si$ (blue)<br>Mg (pale gray) | 100 mL distilled water<br>100 mL ethanol 96%<br>5 g picric acid | Up to 30 s<br>!!! See section 4.4 in the Appendix. |
| Mg m10 | Mg-Zn alloys. Mg-Zn is heavily attacked; $Mg_7Zn_3$, lightly attacked. | 1000 mL distilled water<br>50 g chromium (VI) oxide<br>4 g sodium sulfate | 2 s<br>!!! See section 4.4 in the Appendix. |
| Mg m11 | Mg alloys (forged)<br>Grain-contrast etchant | 100 mL distilled water<br>10 g ammonium persulfate | Swabbing until the surface appears brown |
| Mg m12 | Mg alloys | 2 mL distilled water<br>10 mL picric acid in alcohol 6%<br>1 mL phosphoric acid 85% | Seconds to minutes<br>Suitable for polarized light examination<br>!!! See section 4.4 in the Appendix. |

## Electrolytic Etching

| No. | Material | Etchant | Conditions |
|---|---|---|---|
| Mg Em1 | Complex Mg alloys containing Al, Bi, Cd, Zn | 100 mL distilled water<br>10 g sodium hydroxide | 2–4 min. 4 V DC<br>Cu cathode<br>Etch immediately after polishing. |
| Mg Em2 | Pure Mg and Mg-rich alloys | 20 mL distilled water<br>20 mL ethanol 96%<br>40 mL phosphoric acid 85% | 1–10 min. 10–35 V DC<br>Mg cathode<br>!!! See section 4.4 in the Appendix. |

# *1.9.15 Manganese*

Mn

## Preparation

Pure manganese is used to deoxidize and desulfurize steels and as a catalyst to produce super-purity nitrogen.

Mn ore (ferromanganese) used for the production of Mn steel.
Mn-Cu, Mn-Ni, Mn-Cu-Ni, Mn-Co, and Mn-In alloys.

*Grinding:*

Wet grinding on SiC papers down to 1200 grit.

*Polishing:*

a. Polish with diamond to its finest particle size (0.25 μm) or with 0.05 μm alumina slurry on a short-nap synthetic cloth.
b. Pure Mn and Mn-Cu alloys can be polished electrolytically with Mn EP1.

## Electrochemical Polishing Recipes

*Electrolytic Polishing*

| | | |
|---|---|---|
| Mn EP1 | 50 mL ethanol 96%<br>25 mL glycerol 87%<br>25 mL phosphoric acid 85% | Seconds to minutes<br>18 V DC<br>Stainless steel cathode<br>!!! See section 4.4 in the Appendix. |

## Microetching

| No. | Material | Etchant | Conditions |
|---|---|---|---|
| Mn m1 | Mn-Co, Mn-Cu, Mn-Fe, Mn-Ni alloys<br>High-alloy Mn | 98 mL ethanol 96%<br>2 mL nitric acid 65%<br>(Nital etchant) | Seconds to minutes |
| Mn m2 | Mn-Si-Ca, Mn-Si-Cr, and Mn-Fe alloys | 90 mL distilled water<br>10 mL hydrofluoric acid 40% | Seconds to minutes<br>!!! See section 4.4 in the Appendix. |
| Mn m3 | Mn-Ge, Mn-Si, Mn-Sn-Ge, and Mn-Sn-Si alloys | 40 mL glycerol 87%<br>30 mL hydrofluoric acid 40%<br>25 mL hydrochloric acid 32%<br>10 mL nitric acid 65% | 1–3 s for Mn-Ge and Mn-Si alloys<br>5–10 s for Mn-Sn-Ge and Mn-Sn-Si alloys<br>!!! See section 4.4 in the Appendix.<br>!Explosive! |
| Mn m4 | Mn-Ti alloys (up to 30 Ti) | 60 mL glycerol 87%<br>20 mL nitric acid 65%<br>20 mL hydrofluoric acid 40% | Seconds to minutes<br>!!! See section 4.4 in the Appendix. |
| Mn m5 | Pure Mn. Mn with Cu and Mn with small additions of Ni, Co, Fe, Ge | 200 mL acetyl acetone<br>1–2 mL nitric acid 65% | 2–18 min<br>If possible with ultrasonics to break up the oxide layer |

Ni

## 1.9.16 Nickel

### Preparation

Nickel is a skin and respiratory toxin and may cause allergies. (!!! See section 4.4 in the Appendix.)

I. Pure nickel with purity of 98.5 to 99.8 Carbonyl nickel. Nickel 200 with a purity of 99.5, electrolyte nickel with 99.5 to 99.8 purity; Ni is used for galvanizing and plating.

Pure nickel at 90.0 to 99.5 Ni is used for laboratory equipment.

Pure nickel is tough and tends to form smear and deformation layers.

*Grinding:*

Wet grinding on SiC papers down to 1200 grit.

*Polishing:*

a. Polish with 6 to 1 μm diamond particle size. Remove the deformation layer by etching with Ni EP2 or Ni CP1 between polishing. Final polish with 0.05 μm alumina slurry or with $SiO_2$ suspension.
b. Polish with 6 μm diamond particle size, then on a vibratory polisher for 2 to 3 h with 0.05 μm alumina slurry to which some drops of aqueous ammonium persulfate solution (10%) have been added.
c. After fine grinding, polish with 1 μm chromium oxide on a vibratory polisher for 1 to 2 h. Final polish with $SiO_2$ suspension.
d. Polish electrolytically with Ni EP1, Ni EP2, Ni EP3, Ni EP4, or with Ni EW2.

II. Ni alloys:

- Corrosion-resistant alloys: Ni-30Cu-Fe with Al and Si. Monel alloys with 66.5Ni-31.5Cu-1.25Fe.
- Oxidation- and elevated-temperature-resistant Ni alloys: Ni-20Cr. Inconel with 20 to 70 Ni, remainder is Cr and Fe.
- Magnetic materials: Permalloy Ni-21.5Fe. Superalloy with 36 to 80 Ni and additions of Fe, Cu, Cr, Mn, Mo, and Co. Hard magnets of FeNiCo with additions of Ti, Al, and Nb.
- Material for electrical resistors: Constantan 45Ni-Cu. Ni-Fe-Cr alloys.
- Heat conductor wire for electrical furnaces: Ni-80Cr and Ni-60Cr.
- Heat-resistant materials: Ni-20Cr-Ti, Ni-15Cr-Fe.

*Grinding:*

Wet grinding on SiC papers down to 1200 grit with little pressure.

*Polishing:*

a. Polish with 6 to 0.25 μm diamond particle size. Final polish with 0.05 μm alumina slurry or with $SiO_2$ suspension.
b. Polish with 6 μm diamond particle size, then use vibratory polish for 1 h with 0.05 μm alumina slurry and with some drops of a 10% aqueous ammonium persulfate solution.
c. Polish down to 1 μm diamond particle size, then use electrolytic shock polishing with Ni EP4, 5 $A/cm^2$ for 1 s.

III. High-temperature-resistant Ni-base alloys (superalloys):

- Solid-solution hardenable alloys, e.g., Hastelloy with Ni-16Cr-17Mo-5Fe-0.07C-2.5Co; foundry and wrought alloys.
- Precipitation-hardening alloys, e.g., René 95 with Ni-14Cr-3.5Mo-8Co-2.5Ti-13.5Al-3.5W-3.5Nb, cast material and rolling stock.
- Fe-Cr-Ni alloys with Fe, 37 to 41 Ni, 17 to 21 Cr, 0.3 to 0.7 C, with additions of Mn, Si, Mo, P, and S.
- 26Fe-15Ni-0.05Cr-C with Ti, Mo, V, and Al.

Since most of these materials have a high hardness, automatic sample preparation is recommended.

*Grinding:*

a. Wet grinding on SiC papers down to 1200 grit.
b. Grind flat, then lap with 6 μm diamond particle size on a Bakelite composite lapping disk.

*Polishing:*

a. After grinding of III.a, polish with 15 and 3 μm diamond particle size on a hard synthetic cloth; final polish with 1 and 0.25 μm diamond on a cotton and velvet cloth.
b. After grinding of III.b, polish with 3 μm diamond particle size on a hard synthetic cloth; final polish with 1 and 0.25 μm diamond particle size on a cotton and velvet cloth.
c. After 1 μm diamond polish, shock polish with electrolyte Ni EP4, 5 $A/cm^2$ for 1 to 3 s.

IV. Oxide-dispersion-strengthened (ODS) superalloys.

The matrix of these alloys contains very small metallic oxides, e.g., $ThO_2$. ODS alloys are produced by a sintering process. They are very hard and should only be prepared with automatic equipment.

*Grinding:*

a. On diamond disks down to 20 μm particle size.
b. Flat grinding, then lapping with 6 μm diamond particle size.

*Polishing:*

Polish with 3 and 1 μm diamond particle size on a hard silk cloth.

Final polish with 0.3 μm alumina slurry.

## Electrochemical Polishing Recipes

*Electrolytic Polishing*

| | | |
|---|---|---|
| Ni<br>EP1 | 700 mL glacial acetic acid<br>300 mL perchloric acid 60% | 1–5 min<br>40–100 V DC<br>Ni cathode |
| Ni<br>EP2 | 400 mL distilled water<br>600 mL sulfuric acid 95–97% | 2–6 min<br>1.5–6 V DC<br>Ni cathode |
| Ni<br>EP3 | 660 mL methanol 99.8%<br>330 mL nitric acid 65% | 10–60 s<br>40–70 V DC<br>Ni cathode<br>Do not store!<br>!!! See section 4.4 in the Appendix. |
| Ni<br>EP4 | 200 mL perchloric acid 60%<br>700 mL ethanol 96%<br>100 mL butyl cellosolve 99% | Seconds to minutes<br>30–60 V DC<br>Stainless steel cathode<br>!!! See section 4.4 in the Appendix. |

*Electrolytic Lapping*

| | | |
|---|---|---|
| Ni<br>EW1 | 840 mL distilled water<br>50 g potassium cyanide<br>50 g sodium hydrogen phosphate<br>60 g sodium fluoride | Approximately 5 min<br>1.3 A/0.4 $cm^2$. AC current<br>!!! See section 4.4 in the Appendix. |
| Ni<br>EW2 | 985 mL distilled water<br>12 g sodium thiosulfate<br>1 g copper (II) nitrate<br>2 g thiourea | 3–4 min<br>4–6 mA/0.7 $cm^2$<br>Ni cathode. Can be used with or without fine alumina |
| Ni<br>EW3 | 100 mL distilled water<br>10 mL alumina<br>5 g ammonium chloride<br>2 g potassium nitrate | 0.5–1 min<br>0.05 μm<br>0.05–0.1 mA/0.08 $cm^2$<br>Ni-10Cr cathode |

*Chemical Polishing*

| | | |
|---|---|---|
| Ni<br>CP1 | 30 mL nitric acid 65%<br>10 mL sulfuric acid 95–97%<br>10 mL phosphoric acid 85%<br>50 mL glacial acetic acid | 30–60 s<br>80–90 °C<br>!!! See section 4.4 in the Appendix. |

## Macroetching

Samples should be wet ground on SiC papers down to 1000 grit before macroetching. At reduced etching times, all macroetchants can also be used as microetchants.

| No. | Material | Etchant | Conditions |
|---|---|---|---|
| Ni M1 | High-alloy Ni. Ni-Cu alloys (Monel)<br>Ni-Al and Ni-Fe alloys<br>Porosity, deformation, flow lines | 50 mL distilled water<br>50–100 mL nitric acid 65%<br>(Concentration variable) | 20–30 min |
| Ni M2 | For grain-size measurements in superalloys<br>Ni and Ni-base alloys. Ni-Cu and Ni-Cr-Fe alloys | 50 mL distilled water<br>50 mL ethanol 98%<br>50 mL hydrochloric acid 32%<br>10 g copper (II) sulfate | Seconds to minutes |
| Ni M3 | Low-alloy Ni<br>Cracks and porosity | 10 (100) mL distilled water<br>20 (100) mL nitric acid 65%<br>10 (20) g copper (II) sulfate<br>(Concentration variable) | 20–30 min |
| Ni M4 | Cr- and Fe-containing Ni alloys<br>Welds | 50 mL glacial acetic acid<br>50 mL nitric acid 65%<br>(75) mL hydrochloric acid 32% | Seconds to minutes<br>Swabbing. Hydrochloric acid may be omitted. |
| Ni M5 | Ni-Cr- and Ni-Fe-Cr-base Ni alloys, Inconel type<br>Ni-Nb, Ni-Ta, Ni-Si, Ni-Au, and Ni-Co-Cr alloys | 20–30 mL distilled water<br>0–20 mL nitric acid 65%<br>20 mL hydrochloric acid 32%<br>10 mL hydrogen peroxide 30%<br>(Concentration variable) | 2 min<br>Use only fresh! |
| Ni M6 | Superalloys | 125 mL saturated aqueous iron (III) chloride solution<br>600 mL hydrochloric acid 32%<br>18 mL nitric acid 65% | 5–10 min<br>Use boiling. |
| Ni M7 | Ni-base foundry alloys | 25 mL distilled water<br>50 mL hydrochloric acid<br>15 g iron (III) chloride<br>3 g ammonium tetrachlorocuprate II<br>(Adler's reagent) | min to h |

## Microetching

| No. | Material | Etchant | Conditions |
|---|---|---|---|
| Ni m1 | Pure Ni materials and high Ni-containing alloys<br>Ni-Ti and Ni-Cu alloys<br>Reveals grain boundaries | 0 (10) mL distilled water<br>50 (38) mL nitric acid 65%<br>50 (100) mL glacial acetic acid<br>(Concentration variable) | 5–30 s<br>Use directly after mixing, do not store! |
| Ni m2 | Ni and Ni-base alloys<br>Ni-Cr alloys | 80 mL nitric acid 65%<br>3 mL hydrofluoric acid 40% | Seconds to minutes<br>Preheat sample in boiling water before etching.<br>!!! See section 4.4 in the Appendix. |
| Ni m3 | Pure Ni<br>Superalloys. Ni-Cu and Ni-Fe alloys | 50 mL distilled water<br>10 g copper sulfate<br>50 mL hydrochloric acid 32% | 5–10 s |

(continued)

## Microetching (continued)

| No. | Material | | Etchant | Conditions |
|---|---|---|---|---|
| Ni m4 | Ni-Fe, Ni-Cu, and Ni-Ag alloys<br>Ni-base superalloys<br>Monel | 20–100 mL<br>2–25 mL<br>5–8 g | distilled water or ethanol 96%<br>hydrochloric acid 32%<br>iron (II) chloride<br>(Concentration variable)<br>(Carapella's reagent) | 5–60 s<br>Swabbing |
| Ni m5 | Pure Ni<br><br>Ni-Zn-Ag, Ni-Ag, Ni-Cu, and Ni-Al-Mo alloys<br>Nonmetallic inclusions are not attacked | (a) 100 mL<br>10 g<br>(b) 100 mL<br>10 g | distilled water<br>ammonium persulfate<br>distilled water<br>potassium cyanide<br>1:1 mixture<br>Add a few drops of 3% aqueous hydrogen peroxide solution. | 30–60 s<br>Swabbing<br><br>Seconds to minutes<br>Use fresh only!<br>!!! See section 4.4 in the Appendix. |
| Ni m6 | Ni-Al, Mo-Ni, and Ni-Ti alloys | 100 mL<br>0.1–1 g | hydrochloric acid 32%<br>chromium (VI) oxide | Seconds to minutes<br>!!! See section 4.4 in the Appendix. |
| Ni m7 | Ni-Fe alloys<br>Reveals grain boundaries | | Saturated aqueous ammonium persulfate solution | Seconds to minutes |
| Ni m8 | Ni-Zn alloys | 85 mL<br>5 mL | ammonium hydroxide 25%<br>hydrogen peroxide 30% | 5–15 s<br>Use fresh only! |
| Ni m9 | Ni-Cu alloys and Ni-base superalloys<br>Dissolves protective layers; gamma prime phase is dissolved.<br>Etches very uniformly<br>Suitable for optical microscopy; not suitable for contrasting of the eta phase | 40–80 mL<br>40 mL<br>2 g | ethanol 96%<br>hydrochloric acid 32%<br>copper (II) chloride<br>(Kalling's reagent 2) | Seconds to minutes<br>Immersion or swabbing |
| Ni m10 | Superalloys, Hastelloy type | 50 mL<br>150 mL<br>25 g | distilled water<br>hydrochloric acid 32%<br>chromium (VI) oxide | 5–20 s<br>!!! See section 4.4 in the Appendix. |
| Ni m11 | Ni-Fe and Ni-Al alloys | 40 mL<br>30 mL<br>10 mL<br>20 mL | hydrochloric acid 32%<br>nitric acid 65%<br>glycerol 87%<br>glacial acetic acid<br>(Concentration variable) | Seconds to minutes<br>Add nitric acid at the end.<br>Do not store!<br>!!! See section 4.4 in the Appendix.<br>!Explosive! |
| Ni m12 | Ni silicide | 80 mL<br>10 mL<br>10 mL | distilled water<br>nitric acid 65%<br>hydrofluoric acid 40% | Seconds to minutes<br>!!! See section 4.4 in the Appendix. |
| Ni m13 | Ni-20Cr and Ni-20Cr-45Fe | 50 mL<br>50 mL | distilled water<br>nitric acid 65% | 30–60 s<br>90–100 °C |
| Ni m14 | Inconel X550, Inco 700, Waspaloy, M252<br>Grain boundaries in solid solution. Carbides are heavily attacked. | 97 mL<br>2 mL<br>1 mL | hydrochloric acid 32%<br>sulfuric acid 95–97%<br>nitric acid 65% | 30–120 s |
| Ni m15 | Superalloys<br>Etches gamma phase, carbides<br>Grain-contrast etchant | 100 mL<br>2–4 mL | hydrochloric acid 32%<br>hydrogen peroxide 30% | 10–15 s |

(continued)

## Microetching (continued)

| No. | Material | Etchant | Conditions |
|---|---|---|---|
| Ni m16 | Superalloys<br>Etches gamma phase, carbides, and grain boundaries | 100 mL hydrochloric acid 32%<br>0.4–1 g sodium peroxide | 10–50 s |
| Ni m17 | Ni-Ti alloys<br>Superalloys<br>Etches Al and Al/Cr coatings<br>Dissolves gamma phase. Suitable for SEM examination | 150 mL distilled water<br>25 mL nitric acid 65%<br>10 mL hydrofluoric acid 40% | 5–30 s<br>Swabbing.<br>Let etchant stand for 2–3 h after mixing before use.<br>!!! See section 4.4 in the Appendix. |
| Ni m18 | Ni-base casting alloys<br>Reveals segregations, carbides, and precipitates<br>Wrought alloys: reveals rolling texture, grain-boundary segregations, deformation bands. Welds. Phi phase<br>Color etchant | 100 mL Beraha III stock solution<br>1 g potassium disulfite<br>(Beraha's reagent III)<br><br>**Stock solution Beraha III:**<br>600 mL distilled water<br>50 g ammonium hydrogen fluoride<br>400 mL hydrochloric acid 32% | 30 s to 5 min<br>Wet sample before etching.<br>Can be stored 1–2 h<br><br>Keep stock solution in a plastic bottle.<br>!!! See section 4.4 in the Appendix. |
| Ni m19 | Ni-base alloys with more than 9Mo<br>Hastelloy X. Hastelloy C<br>Color etchant | 50 mL distilled water<br>7 g ammonium hydrogen fluoride<br>50 mL hydrochloric acid 32%<br>0.5 g potassium disulfite<br>(Beraha's reagent III 1) | 5–10 min<br>30–40 °C<br>Wet sample before etching.<br>Can be stored for 1–2 h<br>!!! See section 4.4 in the Appendix. |
| Ni m20 | Superalloys, Hastelloy A type | 50 mL distilled water<br>1.5 g iron (III) chloride<br>50 mL hydrochloric acid 32%<br>0.5 g potassium disulfite | 5–10 min<br>RT<br>Wet sample before etching. |
| Ni m21 | Gamma-prime-age hardening Ni-base alloys<br>Grain-boundary etch | 90 mL distilled water<br>10 mL hydrofluoric acid 40%<br>Boil this mixture, then add molybdic acid to saturation. | Seconds to minutes<br>Wet sample before etching.<br>!!! See section 4.4 in the Appendix. |
| Ni m22 | Alitized surface layers on Ni-base alloys. Plasma coatings. NiAl intermetallic phases<br>Grain-contrast etchant | 95 mL distilled water<br>5 mL hydrofluoric acid 40%<br>Boil this mixture and add molybdic acid to saturation. | Seconds to minutes<br>Wet sample before etching.<br>!!! See section 4.4 in the Appendix. |
| Ni m23 | Ni brazing alloys when brazed to steels<br>Galvanic Ni coatings | 30 (50) mL distilled water<br>70 (50) mL Adler etch solution<br>1–2 g potassium disulfite<br><br>**Adler's etch solution:**<br>25 mL distilled water<br>50 mL hydrochloric acid 32%<br>15 g iron (III) chloride<br>3 g ammonium tetrachlorocuprate II | 5 s to 2 min<br>Do not wet sample before etching. |
| Ni m24 | Superalloys<br>Dissolves gamma prime<br>Grain boundaries and carbides are attacked. | 50 mL distilled water<br>50 mL hydrochloric acid 32%<br>5 mL nitric acid 65%<br>0.2 mL Vogel's etch | 10 min maximum |

## Electrolytic Etching

| No. | Material | Etchant | Conditions |
|---|---|---|---|
| Ni Em1 | Ni and Ni-base alloys<br>Ni-Cu and Ni-Cr alloys<br>Carbide inclusions<br>Grain-boundary etchant | 100 mL distilled water<br>2–50 mL sulfuric acid 95–97% | 5–30 s<br>6 V DC<br>Pt or stainless steel cathode |
| Ni Em2 | Inconel. Ni-Au, Ni-Mo, Ni-Cr alloys<br>Microinhomogeneities in superalloys. Ni-solders | 100 mL distilled water<br>10 g oxalic acid | 10–15 s<br>6 V DC<br>Pt or stainless steel cathode<br>!!! See section 4.4 in the Appendix. |
| Ni Em3 | Ni and Ni alloys<br>Ni-Ag, Ni-Al, Ni-Cr, Ni-Cu, Ni-Fe, and Ni-Ti alloys<br>Grain-contrast etchant | 85 mL distilled water<br>10 mL nitric acid<br>5 mL glacial acetic acid | 20–60 s<br>1.5 V DC<br>Pt or stainless steel cathode<br>Do not store solution! |
| Ni Em4 | Superalloys<br>Attacks eta phase; gamma and gamma-prime phase remain unattacked.<br>Austenite is attacked. | 12 mL phosphoric acid 85%<br>41 mL nitric acid 65%<br>47 mL sulfuric acid 95–97% | Seconds to minutes<br>6 V DC<br>Pt or stainless steel cathode<br>!!! See section 4.4 in the Appendix. |
| Ni Em5 | Ni-base superalloys<br>Gamma precipitates<br>Ti and Nb microsegregations | 85 mL phosphoric acid 85%<br>5 mL sulfuric acid 95–97%<br>8 g chromium (VI) oxide | 5–30 s<br>10 V DC<br>Pt cathode<br>!!! See section 4.4 in the Appendix. |
| Ni Em6 | Carbides in Ni-Cr alloys | 100 mL distilled water<br>10 g potassium cyanide | 3 min<br>6 V DC<br>Pt cathode<br>!!! See section 4.4 in the Appendix. |

## Physical Etching

Carbides in superalloys and intermetallic phases in heat-resistant Ni solders can be identified by interference layer coating. Coating material is ZnSe or a Fe cathode in an oxygen atmosphere (see section 1.7.3 in this chapter).

### *1.9.17 Lead*

Pb

## Preparation

Lead is toxic when inhaled and ingested (!!! See section 4.4 in the Appendix.).

I. Pure lead. Low-alloy lead with less than 0.08 Ca and/or Cu.

Pb is used for shielding of x-rays or radioactive radiation and also for protective tubing of cables.

Pb is extremely soft and ductile, as well as temperature sensitive. It tarnishes when exposed to water. Cold deformation and recrystallization cannot be avoided during cutting and grinding. However, it must be minimized by careful sample preparation and then removed by the subsequent polishing steps. Smear layers and ghost structures can even be formed during polishing. Grit particles and polishing compounds embed easily when working with too much pressure.

*Sectioning:*

Section with a handsaw or a low speed cutoff wheel such as a Bakelite-bonded SiC cutoff wheel with water or methanol as a coolant.

*Mounting:*

Do not use clamps. Mount with a room-temperature curing compound that has a low curing temperature. If necessary, cool the mount.

*Grinding:*

a. Ultramilling (micromilling) will produce only slight deformation and a surface tarnish is avoided.
b. Grind manually on wax-treated SiC papers down to 1200 grit. The sample is moved in a unidirectional upward motion only, to avoid the embedding of SiC grit particles. Use low pressure.

*Polishing:*

a. After ultramilling polish with a $SiO_2$ suspension on a short-nap velvet cloth if necessary.
b. Polish with 9 μm diamond particle size on a cotton cloth on a rotating disk, changing direction of the sample at 90° several times; use an oil-based lubricant. Follow with rotational or vibratory polishing on a short-nap velvet cloth with 0.3 μm alumina to which a few drops of hydrogen peroxide and glacial acetic acid have been added. Another good polishing compound is cerium oxide.
c. Remove the deformed layer after grinding by etching with Pb M3; then polish with 1 μm alumina on a felt cloth, or with 3 and 1 μm diamond particle size on a cotton cloth using a lubricant and few drops of hydrogen peroxide until the microstructure becomes visible. The final polishing is carried out on a short-nap velvet cloth with a $SiO_2$ suspension or with 0.05 μm alumina (use distilled water for the slurry) with a few drops of ammonium tartrate. Rinse the prepared samples only in distilled water or alcohol. If the sample surface appears slightly dull, submerse briefly into a diluted solution of Villela's reagent (Pb M5).
d. Electrolytic polish with the electrolyte Pb EP1.

II. Soft solders: Pb-Sn-base alloys, with or without Ag.

*Grinding:*

Wet grinding on a rotating wheel with SiC papers down to 1200 grit, then continue on microcut paper. Change the direction frequently, do not use any pressure; long grinding times are necessary.

*Polishing:*

Polish with 6 to 1 μm diamond particle size on a cotton cloth, using an oil-based lubricant.

III. Hard lead: Pb-Sb for batteries (accumulators).

Lead-bearing alloys: Pb-Sn-Sb with Cu, As, and Cd.

Type metal: 80Pb-5Sn-15Sb with or without Cu and As.

The preparation of these materials is not too difficult. The soft and mostly eutectic matrix contains hard phases (e.g., SnSb, $Cu_6Sn_5$).

*Grinding:*

Wet grinding on SiC papers down to 1200 grit; use low speed and little pressure.

*Polishing:*

Polish with 6 μm diamond particle size on a hard silk cloth using an alcohol-based lubricant. Final polish with 0.3 μm alumina slurry to which ammonium tartrate has been added (1 g per 1 L distilled water).

## Electrochemical Polishing Recipes

*Electrolytic Polishing*

| | | |
|---|---|---|
| Pb<br>EP1 | 300 mL perchloric acid 60%<br>700 mL glacial acetic acid,add in drops, cooling the solution with ice! | 3–4 min. 15 °C<br>8 V DC<br>Stainless steel cathode<br>!!! See section 4.4 in the Appendix. |
| Pb<br>EP2 | 700 mL ethanol 96%<br>100 mL glycerol 87%<br>200 mL perchloric acid 60% | 15–60 s at 20 °C<br>15–50 V DC<br>Stainless steel cathode<br>!!! See section 4.4 in the Appendix. |

## Macroetching

| No. | Material | Etchant | Conditions |
|---|---|---|---|
| Pb M1 | Pure Pb. Coatings. Welds<br>Grain-contrast etchant | 80 mL distilled water<br>20 mL nitric acid 65%<br>(Concentration variable) | 10 min<br>Swabbing |
| Pb M2 | Pure Pb and low-alloy Pb<br>Pb-Sb alloys<br>Removes deformed layers | (a) 100 mL distilled water<br>15 g ammonium molybdate<br>(b) 42 mL distilled water<br>58 mL nitric acid 65% | 10–30 s<br>Mix equal parts of (a) and (b).<br>Swab the samples. Polish and etch alternately, rinsing in running water. |
| Pb M3 | Pure Pb. Pb-Ca and Pb alloys with low Sn<br>Sb containing Pb<br>Grain-contrast etchant | 68 mL glycerol 87%<br>16 mL glacial acetic acid<br>16 mL nitric acid 65% | Seconds to minutes<br>80 °C. Suitable also as a microetchant<br>Use fresh only!<br>!!! See section 4.4 in the Appendix.<br>!Explosive! |
| Pb M4 | Pb-Sb alloys<br>Grain-contrast etchant | 100 mL distilled water<br>25 g citric acid<br>10 g ammonium molybdate | Seconds to minutes |
| Pb M5 | Pb-Sb alloys<br>Grain-boundary etchant | 80 mL glycerol 87%<br>10 mL glacial acetic acid<br>10 mL nitric acid 65%<br>(Villela's and Beregekoff's reagent) | Seconds to minutes. 80 °C<br>Polish and etch alternately.<br>!!! See section 4.4 in the Appendix.<br>!Explosive! |
| Pb M6 | Pb-Sb alloys<br>Grain-contrast etchant | 80 mL glacial acetic acid<br>20 mL hydrogen peroxide 30% | Few s |
| Pb M7 | Pb-Sb-Cu alloys<br>Grain-contrast etchant | 95 mL ethanol 96%<br>5 mL glacial acetic acid | 20 min |
| Pb M8 | Pb-As-Sn-Bi alloys<br>Grain-contrast etchant | 100 mL distilled water<br>75 mL nitric acid 65%<br>20 mL ammonium hydroxide solution 25%<br>16 g ammonium molybdate | Seconds to minutes |
| Pb M9 | Pb alloys | (a) 25 mL distilled water<br>10 mL saturated (50 °C) aqueous ammonium molybdate solution<br>14 mL ammonium hydroxide solution 25%<br>(b) 96 mL distilled water<br>40 mL nitric acid 65% | Seconds to minutes<br>Filter (a) then add (b). |

## Microetching

| No. | Material | Etchant | Conditions |
|---|---|---|---|
| Pb m1 | Pure Pb. Pb-Ca, Pb-Sb, Pb-Cd alloys. Pb containing Cu and bronzes. High-alloy Pb white metals. Segregations<br>Grain-boundary etchant | 84 mL glycerol 87% or ethanol 96%<br>8 mL glacial acetic acid<br>8 mL nitric acid 65% | Few s at 80 °C<br>Use fresh only!<br>!!! See section 4.4 in the Appendix.<br>!Explosive! |
| Pb m2 | Pure Pb. Pb-Ca alloys<br>Pb with less than 2% Sb | 60 mL distilled water<br>15 mL glacial acetic acid<br>5 mL hydrogen peroxide 30% | 20 s |

(continued)

## Microetching (continued)

| No. | Material | Etchant | Conditions |
|---|---|---|---|
| Pb m3 | Pure Pb. Pb-Ca and Pb-Cu alloys | 60 mL glycerol 87%<br>15 mL nitric acid 65%<br>15 mL glacial acetic acid | Seconds to minutes. 80 °C<br>Use only fresh!<br>!!! See section 4.4 in the Appendix.<br>!Explosive! |
| Pb m4 | Pure Pb. Pb-Na alloys<br>Grain boundary etching | 30 mL glacial acetic acid<br>10 mL hydrogen peroxide 30% | 8–15 min |
| Pb m5 | Pure Pb<br>Grain boundaries and grain-contrast etchant | 100 mL distilled water<br>7 mL nitric acid 65% | 10–25 s |
| Pb m6 | Alpha phase in Pb-Bi and Pb-Sb-Sn alloys | 50 mL lactic acid 90%<br>30 mL nitric acid 65%<br>20 mL hydrogen peroxide 30% | 30 s to 1 min<br>Swabbing |
| Pb m7 | Epsilon phase in Pb-Sb-Sn alloys | 50 mL lactic acid 90%<br>10 mL nitric acid 65%<br>7 mL hydrogen peroxide 30% | Seconds to minutes<br>Swabbing<br>Alpha etches before epsilon. |
| Pb m8 | Pb-Ca alloys<br>Inclusions<br>Grain-boundary etchant | 60 mL glacial acetic acid<br>30 mL hydrogen peroxide 30% | 8–15 s |
| Pb m9 | Pb-Ca alloys | (a) 20 mL glacial acetic acid<br>10 mL hydrogen peroxide 30%<br>(b) 50 mL distilled water<br>50 mL nitric acid 65% | Seconds to minutes<br>Time depends on the thickness of the deformed layer.<br>Etch first in (a) then in (b). |
| Pb m10 | Pb-Cd and Pb-Sn alloys<br>Precipitates<br>Grain-boundary etchant | 76 mL glycerol (or ethanol 96%)<br>16 mL glacial acetic acid<br>8 mL nitric acid 65%<br>(Concentration variable) | Seconds to minutes. If necessary at 80 °C<br>Use only fresh!<br>!!! See section 4.4 in the Appendix.<br>!Explosive! |
| Pb m11 | Pb-Cd, Pb-Sn, and Pb-Sb alloys | 100 mL distilled water<br>25 mL nitric acid 65%<br>16 mL glacial acetic acid | 4–30 min. 40 °C<br>Rinse with cotton under running water. |
| Pb m12 | Pb-Sb alloys. Hard lead<br>White metal and type metal with high Pb content | 100 mL ethanol 96%<br>1–5 mL nitric acid 65% | 1–10 min. After etching, immerse briefly in tap water. Pb solid-solution phases will discolor dark.<br>Also suitable as a macroetchant |
| Pb m13 | Pb-Sb alloys. Lead solders<br>Pb-containing white and type metal | 90 mL distilled water (or ethanol 96%)<br>30 mL hydrochloric acid 32%<br>10 g iron (III) chloride<br>(Concentration variable) | 1–10 min |
| Pb m14 | All Pb and Pb-Sb alloys<br>Pb-containing white and type metals and bearing alloys | 100 mL distilled water<br>5–10 g silver nitrate 32% | Seconds to minutes<br>Swab etching |
| Pb m15 | Pb-Sb alloys<br>Pb-rich phase: dark<br>Sb-rich phase: light | 100 mL distilled water<br>5 mL nitric acid 65% | 5–40 s |
| Pb m16 | Pb containing less than 2 Sb | 100 mL glacial acetic acid<br>10 mL hydrogen peroxide 30% | 10–30 min |
| Pb m17 | Pb-Sb-Sn alloys<br>Sb-Sn crystals: light<br>$Ca_2Sb$ crystals: light<br>Matrix: dark | 10 mL distilled water<br>40 mL glacial acetic acid<br>20 mL nitric acid 65% | Few s |

## Electrolytic Etching

| No. | Material | Etchant | Conditions |
|---|---|---|---|
| Pb Em1 | Pure Pb, Pb-Sb, and Pb-Sn alloys | 40 mL distilled water<br>60 mL perchloric acid 60% | 10 s. 2 V DC<br>Pt anode<br>The sample is the cathode.<br>!!! See section 4.4 in the Appendix. |
| Pb Em2 | Pb alloys<br>Grain-contrast etchant | 70 mL glacial acetic acid<br>30 mL perchloric acid 60%<br>(Add glacial acetic acid only in drops and cool solution with ice.) | 1 min. 1–2 V DC<br>Pb or Cu cathode<br>!!! See section 4.4 in the Appendix. |

Pt

Pd

Rh

Ir

Ru

Os

### *1.9.18 Platinum Metals (Platinum, Palladium, Rhodium, Iridium, Ruthenium, and Osmium)*

**Preparation**

I. Pure platinum (99.99) is used for pyrometer parts, wires, and electrodes.

Pure palladium (99.94) is used for telephone relays.

Pure Pt and Pd are soft.

*Sectioning:*

Section with a tungsten carbide (hardmetal) cutoff wheel.

*Grinding:*

Wet grinding on SiC papers down to 2400 grit.

*Polishing:*

a. Polish with 6 μm diamond particle size on a hard synthetic cloth, then with 1 μm diamond particle size on a hard silk cloth.
Final polish with 0.25 μm diamond particle size on short-nap synthetic cloth. Preparation of the final stage requires a long time.

b. Final polish Pd with Pt EW1.

II. 75 to 97 Pt and 3 to 25 Ir. 95Pt-5Au. 60 to 90 Pt and 10 to 40 Rh is used for thermocouples and instrumentation.

95Pt-5Ru is used for contacts and electrical resistance wires.

75Pt-23Co is used for magnet material.

92Pt-8W is used for spark plug electrodes.

50Pt-Au with Rh is used for spin jets.

Pd-Cu is used for slide contacts and springs.

93.5Pd and 6.5Ru is used for electrical resistance wire and white jewelry alloys.

Pd-Au (white gold) is used for jewelry.

Pd-Ag-Cu. Pd with Al, Ga, In, Sn, Fe, and Cu are used as dental alloys.

The harder alloys can be prepared as described in I.

*Grinding:*

Wet grinding on SiC papers down to 2400 grit.

*Polishing:*

a. Polish with 3 μm diamond particle size on a cotton cloth and with 1 and 0.25 μm diamond particle size on a short-nap synthetic cloth.
b. Single-phase and two-phase Pd alloys with Pt EW1.
c. Etch polish single-phase Pd alloys with 0.05 μm alumina slurry and a few mL of Pt m3 etchant.
d. Polish hard and multiphase dental alloys on a vibratory polisher with:
   300 mL water
   30 mL of 0.05 μm alumina
   30 mL of aqueous ammonium persulfate solution 20%
   30 mL potassium ferricyanide

III. Pure Ir and pure Rh are used for analytical equipment.

*Grinding:*

Wet grinding on SiC papers down to 2400 grit.

*Polishing:*

Polish with 3 μm diamond particle size on a cotton cloth and then with 1 and 0.25 μm diamond particle size on a short-nap synthetic cloth.

IV. Ru and Os are used as alloying elements. Both metals are very hard.

*Sectioning:*

Section with a diamond cutoff wheel.

*Grinding:*

Wet grinding on SiC papers down to 2400 grit.

*Polishing:*

Polish with 3 μm diamond particle size on a cotton cloth and then with 1 and 0.25 μm diamond particle size on a short-nap synthetic cloth using a high rotating wheel speed. Final polish with 0.05 μm alumina slurry on a vibratory polisher.

## Electrochemical Polishing Recipes

### *Electrolytic Lapping*

| | | |
|---|---|---|
| Pt<br>EW1 | 820 mL distilled water<br>50 g sodium thiosulfate<br>70 g potassium thiocyanate<br>60 g ammonium chloride | Several min<br>1.2–1.5 A/cm² DC on cloth at low speed<br>If the sample stays dull |

## Macroetching

| No. | Material | Etchant | Conditions |
|---|---|---|---|
| Electrolytic:<br>Pt<br>M1 | Pt and Pt alloys | 80 mL saturated aqueous sodium chloride solution<br>20 mL hydrochloric acid 32% | Several min. 6 V DC<br>Pt cathode |

## Microetching

| No. | Material | Etchant | Conditions |
|---|---|---|---|
| Pt<br>m1 | Pure Pt and Pd with low alloy content<br>Rh, Ru, Ir, and Os | 50 mL distilled water<br>100 mL hydrochloric acid 32%<br>10 mL nitric acid 65% | 1–5 min<br>Use hot.<br>!!! See section 4.4 in the Appendix. |
| Pt<br>m2 | Pt and Pd alloys with noble metal content below 90 | (a) 100 mL distilled water<br>10 g potassium cyanide<br>(b) 100 mL distilled water<br>10 g ammonium persulfate | 30 s to 2 min<br>Mix (a) and (b) at a 1:1 ratio.<br>!!! See section 4.4 in the Appendix. |
| Pt<br>m3 | Pd dental alloys<br>Pd alloys with Al, Ga, In, Sn, Fe, and Cu | 30 mL glycerol<br>40 mL nitric acid 65%<br>40 mL hydrochloric acid 32%<br>(Villela's reagent) | 2–10 s<br>Rinse in 5% aqueous hydrochloric acid.<br>Use fresh only!<br>!!! See section 4.4 in the Appendix.<br>!Explosive! |

## Electrolytic Etching

| No. | Material | Etchant | Conditions |
|---|---|---|---|
| Pt<br>Em1 | Ir<br>Rh | 75 mL distilled water<br>25 mL hydrochloric acid 32% | Seconds to minutes<br>3 V AC. Pt cathode |
| Pt<br>Em2 | Pt and Pt alloys<br>Rh and Rh alloys<br>Pd-Ag alloys<br>Ir | 100 mL distilled water<br>10 g potassium cyanide | 1–2 min<br>1–5 V AC. 0.1–0.2 A/cm²<br>Pt or graphite cathode<br>!!! See section 4.4 in the Appendix. |
| Pt<br>Em3 | (1) Pure Pt and Pt alloys<br>(2) Rh-base alloys<br>(3) Pt-10Rh alloys<br>(4) Pt-Ir alloys<br>(5) Ru-base alloys | 65 mL distilled water<br>20 mL hydrochloric acid 32%<br>25 g sodium chloride | (1) 1 min. 6 V AC<br>(2) 25 s. 10 V AC<br>(3) 1 min. 1.5 V AC<br>(4) 1–2 min. 20 V AC<br>(5) 1 min. 5–20 V AC<br>Graphite or Pt cathode |
| Pt<br>Em4 | Pure Ru | 100 mL distilled water<br>10 g oxalic acid | Seconds to minutes<br>AC. Pt or graphite cathode |

(continued)

**Electrolytic Etching (continued)**

| No. | Material | Etchant | Conditions |
|---|---|---|---|
| Pt<br>Em5 | Rh<br>Ru | 50 mL distilled water<br>35 mL hydrochloric acid 32%<br>15 mL hydrogen peroxide 30% | Seconds to minutes<br>3 V AC<br>Pt or graphite cathode |
| Pt<br>Em6 | Os | 100 mL hydrochloric acid 32%<br>25 g sodium chloride | 2–5 min<br>4 V DC<br>Graphite cathode |

## *1.9.19 Plutonium, Thorium, and Uranium, Americium, and Neptunium*

Pu

Th

U

Am

Np

**Preparation**

These metals and their alloys are radioactive and a health hazard. The entire preparation must be done in a glove box that is under lower pressure than the laboratory atmosphere. Etching and the microscopic examination also require special measures to prevent contamination for the operating personnel. This type of metallography can only be carried out in specially equipped laboratories under the supervision of experienced professionals and where the radioactivity is continuously monitored. (!!! See section 4.4 in the Appendix.).

*Grinding:*

Wet grinding on SiC papers down to 1200 grit.

*Polishing:*

a. Polish with alumina slurries or diamond to the finest particle size.
b. Polish electrolytically with Pu EP1 (especially Pu-Al alloys), Pu EP2, and Pu EP3.

**Electrochemical Polishing Recipes**

*Electrolytic Polishing*

| | | |
|---|---|---|
| Pu<br>EP1 | 900 mL glacial acetic acid<br>100 mL perchloric acid 60% | 30 s to minutes. 12–20 V DC<br>Stainless steel cathode<br>!!! See section 4.4 in the Appendix. |
| Pu<br>EP2 | 300 mL ethanol 96%<br>300 mL phosphoric acid 85%<br>300 mL glycerol 87% | 4–6 min. 25–30 V DC<br>Al cathode<br>!!! See section 4.4 in the Appendix. |
| Pu<br>EP3 | 445 mL ethanol 96%<br>275 mL ethylene glycol<br>275 mL phosphoric acid 85% | 1–5 min. 18–20 V DC<br>Stainless steel cathode<br>!!! See section 4.4 in the Appendix. |

## Macroetching

| No. | Material | Etchant | Conditions |
|---|---|---|---|
| Pu M1 | Pure U and U alloys<br>!!! See section 4.4 in the Appendix. | (a) nitric acid 65%<br>(b) hydrochloric acid 32% | 1–5 s<br>Remove the oxidized layer with (a); then treat with (b) until a dull black film forms on the surface; treat again with (a) until sufficient contrast becomes visible. |
| Pu EM1 | Pure U and most U alloys<br>!!! See section 4.4 in the Appendix. | 90 mL distilled water<br>10 mL phosphoric acid 85% | Electrolytic: up to 15 min<br>20 V DC<br>Al cathode |

## Microetching

| No. | Material | Etchant | Conditions |
|---|---|---|---|
| Pu m1 | U alloys (U-N-Mo alloys)<br>!!! See section 4.4 in the Appendix. | Heat tinting | 1–2 h<br>100–200 °C |
| Pu m2 | Pure U and U alloys. U carbides<br>!!! See section 4.4 in the Appendix. | 30 mL nitric acid 65%<br>30 mL glacial acetic acid<br>30 mL glycerol 87% | 5–30 s<br>!!! See section 4.4 in the Appendix.<br>!Explosive! |
| Pu m3 | U-Al alloys:<br>$UAl_2$ (light blue)<br>$UAl_3$ (yellow)<br>$UAl_4$ (gray)<br>!!! See section 4.4 in the Appendix. | 100 (50) mL distilled water<br>38 (25) mL nitric acid 65%<br>or<br>1 (25) mL hydrofluoric acid 40%<br>(Concentration variable) | Seconds to minutes<br>!!! See section 4.4 in the Appendix. |
| Pu m4 | U alloys<br>!!! See section 4.4 in the Appendix. | 70 mL phosphoric acid 85%<br>25 mL sulfuric acid 95–97%<br>5 mL nitric acid 65% | Seconds to minutes |
| Pu m5 | U-Be alloys. Also U beryllides<br>U-Zr and U-Nb alloys. U silicides<br>!!! See section 4.4 in the Appendix. | 30 mL distilled water or lactic acid 90%<br>30 mL nitric acid 65%<br>1 mL hydrofluoric acid 40% | 5–30 s<br>Rinse first in water, then with alcohol.<br>Use lactic acid for U silicides.<br>!!! See section 4.4 in the Appendix. |
| Pu m6 | U-Mo and U-Zr alloys with stabilized gamma phase<br>!!! See section 4.4 in the Appendix. | 40 mL glycerol 87%<br>40 mL nitric acid 65%<br>10 mL hydrofluoric acid 40% | 5–10 s<br>!!! See section 4.4 in the Appendix.<br>!Explosive! |
| Pu m7 | Pure Th and Th alloys<br>!!! See section 4.4 in the Appendix. | 10 mL hydrofluoric acid 40%<br>10 mL nitric acid 65% | Several s<br>!!! See section 4.4 in the Appendix. |
| Pu m8 | U-Zr alloys. U-Fe-Co alloys<br>$U$-$Al_2$-$U_3Si_2$ alloys, $U_3Si_2$ is darker than $UAl_2$.<br>!!! See section 4.4 in the Appendix. | 100 mL distilled water<br>10 g potassium hydroxide<br>10 g potassium ferricyanide<br>(Murakami's reagent) | Seconds to minutes. Use boiling.<br>20–60 s for U-Fe-Co alloys<br>Use only fresh! |
| Pu m9 | U-10Nb<br>!!! See section 4.4 in the Appendix. | 100 mL distilled water<br>5 g chromium (VI) oxide<br>10 drops hydrofluoric acid 40% | Seconds to minutes<br>Add hydrofluoric acid just before use.<br>!!! See section 4.4 in the Appendix. |
| Pu m10 | $U_3Si$. Color etchant<br>!!! See section 4.4 in the Appendix. | 70 mL glacial acetic acid<br>30 mL nitric acid 65% | Seconds to minutes |

(continued)

## Microetching (continued)

| No. | Material | Etchant | Conditions |
|---|---|---|---|
| Pu m11 | U. Pu. Am<br>!!! See section 4.4 in the Appendix. | 75 mL nitric acid 65%<br>25 mL hydrogen peroxide 30% | Seconds to minutes<br>40–50 °C<br>!!! See section 4.4 in the Appendix. |
| Pu m12 | U. Pu<br>!!! See section 4.4 in the Appendix. | 80 mL distilled water<br>20 mL nitric acid 65% | Seconds to minutes<br>50 °C |

## Electrolytic Etching

| No. | Material | Etchant | Conditions |
|---|---|---|---|
| Pu Em1 | Pure U. Pure Th<br>!!! See section 4.4 in the Appendix. | 90 mL glacial acetic acid<br>10 mL perchloric acid 60% | 5–15 min. 18–20 V DC<br>Stainless steel cathode<br>!!! See section 4.4 in the Appendix. |
| Pu Em2 | Pure Pu and Pu alloys<br>!!! See section 4.4 in the Appendix. | 50 mL ethylene glycol<br>20 mL methanol 99.8%<br>5 mL nitric acid 65% | 2 min. 0.05 A/cm$^2$<br>Stainless steel cathode<br>!!! See section 4.4 in the Appendix. |
| Pu Em3 | U-C alloys<br>!!! See section 4.4 in the Appendix. | 6 mL distilled water<br>60 mL glacial acetic acid<br>5 g chromium (VI) oxide | 15 min. 10 V DC<br>Stainless steel cathode<br>!!! See section 4.4 in the Appendix. |

### *1.9.20 Tin*

Sn

## Preparation

I. Pure tin.

Commercial grade Sn of 99.5 to 98 contains impurities of Pb, Sb, Cu, Bi, and Fe.

Sn is used for protective coatings against corrosion and for some decorative parts and musical organs.

Sn deforms very easily and has a low recrystallizing temperature; it tends to form deformation and smear layers easily (this applies also for Sn alloys), as well as ghost structures and deformation twins. It is therefore recommended to use techniques that do not generate heat or mechanical deformation.

*Sectioning:*

The samples can be cut, but a coolant (water or ethanol) must be used to avoid frictional heat.

*Mounting:*

Use cold-setting mounting materials.

Where good edge retention is required, the samples should be plated with copper or nickel before mounting.

*Grinding:*

a. When using microtoming or ultramilling with a diamond knife the polishing step can be omitted.
b. Wet grinding on SiC papers down to 800 grit; the following papers down to 1200 grit should be coated with wax or stearin.

*Polishing:*

Electrolytic polishing with Sn EP1, Sn EP2, Sn EP3.

II. Tin plate: Sn-Fe and Sn-Zn alloys for cans.

Pewter: Sn-Cu used for decorative parts and in art.

Bearing materials: 63 to 80 Sn-Sb-Cu, Sn-18Pb-15Sb-2Cu.

Soft solders: Sn-Sb, Sn-Ag, Sn-Pb alloys.

*Grinding:*

Wet grinding on SiC papers down to 800 grit, the following papers down to 1200 grit should be coated with wax or stearin.

*Polishing:*

a. With 6, 3, 1, and 0.25 μm diamond particle size on a short-nap silk cloth. Double the polishing time for every subsequent step. Final polish on a high-nap cloth with the finest alumina grade or with a suspension of:
   120 mL distilled hot water
   20 mL 5% solution of aqueous ammonium tartrate
   1 g magnesia
   This suspension must be filtered before using.
b. Polish high-Sn-containing materials with 6 and 1 μm diamond particle size and final polish with 0.25 μm diamond using a water-free coolant.
c. Rough polish with 1 μm diamond particle size and final polish electrolytically with Sn EP1, Sn EP2, or Sn EP3.
d. Chemical polish Sn-Te with Sn CP1.

## Electrochemical Polishing Recipes

*Electrolytic Polishing*

| | | |
|---|---|---|
| Sn EP1 | 13 mL distilled water<br>300 mL glacial acetic acid<br>63 mL perchloric acid 60% | 10 min. 20–30 V DC<br>Sn cathode<br>!!! See section 4.4 in the Appendix. |

(continued)

*Electrolytic Polishing (continued)*

| | | |
|---|---|---|
| Sn<br>EP2 | 780 mL distilled water<br>200 mL fluoroboric acid 35%<br>20 mL sulfuric acid 95–97% | 2–5 s. 20–40 °C<br>15–17 V DC<br>Stainless steel cathode<br>!!! See section 4.4 in the Appendix. |
| Sn<br>EP3 | 100 mL distilled water<br>20 g potassium hydroxide | 15 min<br>2–6 V DC<br>Cu cathode |

*Chemical Polishing*

| | | |
|---|---|---|
| Sn<br>CP1 | 0.35 g iodine<br>40 mL ethanol 96%<br>10 mL distilled water<br>4 mL hydrofluoric acid 40% | Mix ingredients in the described sequence.<br>Sample must be polished with 1 µm alumina.<br>Add the solution onto a polishing cloth and move the sample, describing a figure 8.<br>Polish for 15–20 min.<br>!!! See section 4 in the Appendix. |

## Macroetching

*Preparation:*

Coarse (wet) grinding or removal with a special file for light and soft metals is usually sufficient. If necessary, polish briefly with 6 µm diamond particle size.

| No. | Material | Etchant | Conditions |
|---|---|---|---|
| Sn<br>M1 | Sn-rich bearing and tin plate | 100 mL distilled water<br>2 mL hydrochloric acid 32%<br>10 g iron (III) chloride | 30 s to 5 min |
| Sn<br>M2 | Pure Sn | 100 mL distilled water<br>10 g ammonium persulfate | Seconds to minutes |
| Sn<br>M3 | Pure Sn | 100 mL ethanol 96%<br>1–5 mL hydrochloric acid 32% | Seconds to minutes |
| Sn<br>M4 | Sn-Pb alloys | 80 mL glycerol 87%<br>10 mL nitric acid 65%<br>10 mL glacial acetic acid | 1–10 min. 40 °C<br>!!! See section 4.4 in the Appendix.<br>!Explosive! |
| Sn<br>M5 | Sn-Sb-Cu alloys<br>All Sn materials<br>High-Sn-containing alloys | Saturated (cold) aqueous solution of ammonium polysulfide | 20–30 min<br>If necessary wipe the sample surface with cotton after etching. |
| Sn<br>M6 | Alloys with low Sn content<br>High-Cu-containing alloys<br>Grain-contrast etchant | 10 mL hydrogen peroxide 30%<br>10 mL ammonium hydroxide 25% | Seconds to minutes<br>Pre-etch with Sn M5.<br>!!! See section 4.4 in the Appendix. |

## Microetching

| No. | Material | | Etchant | Conditions |
|---|---|---|---|---|
| Sn m1 | Pure Sn. Sn-Cd, Sn-Fe, Sn-Pb, and Sn-Sb alloys | 100 mL<br>2–5 mL | distilled water or ethanol 96%<br>hydrochloric acid 32% | 1–3 min<br>Swabbing |
| Sn m2 | Pure Sn. Sn-rich alloys with Cd, Cu, Fe, and Sb | 100 mL<br>1–5 mL | ethanol 96%<br>nitric acid 65%<br>(Nital reagent) | s to several min<br>If necessary, etch again with a nital reagent of higher concentration. |
| Sn m3 | Pure Sn<br>Sn with 1 Pb | 5 mL<br>3 mL<br>1 mL | glycerol 87%<br>glacial acetic acid<br>nitric acid 65% | To 10 min. 40 °C<br>!!! See section 4.4 in the Appendix.<br>!Explosive! |
| Sn m4 | Pure Sn<br>Sn-Bi alloys | 30 mL<br>60 mL<br>5 mL<br>12 g | distilled water<br>ethanol 96%<br>hydrochloric acid 32%<br>iron (III) chloride<br>Add 1 drop hydrogen peroxide 30% per 100 mL solution. | Seconds to minutes |
| Sn m5 | Sn-rich bearing materials and tin plate<br>Sn-Cu alloys. Sn-Bi eutectic | 100 mL<br>5–25 mL<br>10 g | distilled water or methanol 99.8%<br>hydrochloric acid 32%<br>iron (III) chloride | 10–30 s<br>(If necessary up to 5 min). |
| Sn m6 | Sn-Ca alloys | 50 mL<br>10 mL | glacial acetic acid<br>hydrogen peroxide 30% | Seconds to minutes |
| Sn m7 | Sn-Pb alloys<br>Sn coatings on steel | 25 mL<br>2 mL<br>1 drop | glycerol 87%<br>hydrofluoric acid 40%<br>nitric acid 65% | 1 min<br>!!! See section 4.4 in the Appendix.<br>!Explosive! |
| Sn m8 | Sn-Pb alloys<br>Sn-containing bearing materials<br>Sn coatings | 100 mL<br>5–10 g | distilled water<br>ammonium persulfate | s to several min |
| Sn m9 | Sn-Pb alloys<br>Sn solders | 50 mL<br>50 mL<br>1 drop | distilled water<br>glacial acetic acid<br>hydrogen peroxide 30% | Seconds to minutes |
| Sn m10 | Sn-Pb alloys<br>Pb discolors black.<br>Sn remains unattacked. | 90 mL<br>5–10 g | distilled water<br>silver nitrate | Seconds to minutes<br>Swabbing |
| Sn m11 | Sn-Sb alloys<br>Sn-rich phase discolors black, Sb-rich phase remains light, and SnSb crystals turn yellowish. | (a) 100 mL<br>200 g<br>2 g<br>1.5 g | distilled water<br>chromium (VI) oxide<br>sodium fluoride<br>sodium sulfate | 2 s<br>!!! See section 4.4 in the Appendix. |
| | | (b) 100 mL<br>5 g<br>0.4 g | distilled water<br>chromium (VI) oxide<br>sodium sulfate | 1 s<br>Etch first in (a) and then in (b).<br>!!! See section 4.4 in the Appendix. |
| Sn m12 | Sn-Sb-Cd alloys<br>Epsilon phase discolors brown; delta phase remains unattacked. | 100 mL<br>10 g<br>10 g | distilled water<br>sodium hydroxide<br>potassium ferricyanide | Several min<br>Use boiling. |
| Sn m13 | Sn coatings on steels and cast iron | 100 mL<br>4 g | ethanol 96%<br>picric acid<br>(Picral reagent) | Seconds to minutes<br>!!! See section 4.4 in the Appendix. |
| Sn m14 | Pure Sn. Sn alloys<br>Grain-contrast etchant<br>Color etchant | 50 mL<br>5 g | saturated (cold) sodium thiosulfate<br>potassium disulfite | 60–90 s |

## Electrolytic Etching

| No. | Material | Etchant | Conditions |
|---|---|---|---|
| Sn Em1 | Pure Sn | 80 mL distilled water<br>20 mL sulfuric acid 95–97% | Seconds to minutes. 30 V DC<br>Al cathode<br>When finished, leave current on, while removing the sample. |
| Sn Em2 | Pure Sn<br>Sn-rich alloys | 25 mL distilled water<br>300 mL glacial acetic acid<br>50 mL perchloric acid 60% | 10 min. 20–30 V DC<br>Sn cathode<br>Mix carefully; the solution must be kept cooled!<br>!!! See section 4.4 in the Appendix. |

Ti

## *1.9.21 Titanium*

### Preparation

I. Pure titanium, technical grade titanium: 99.8 to 99.5. Ti is used in equipment for the chemical industry and for galvanotechnical applications.

Unalloyed Ti is soft and tough and tends to form excessive deformation layers.

*Sectioning:*

Use a corundum cutoff wheel at very low speed and with ample coolant.

*Mounting:*

Choose a chemical-resistant mounting compound because of highly aggressive etching solutions.

*Grinding:*

a. When microtoming with a tungsten carbide knife, only a final polishing step will be necessary.
b. When wet grinding on SiC papers down to 2400 grit (if necessary to 4000 grit), use low pressure and long times on the finer grit papers to remove deformation.

*Polishing:*

a. After grinding of (b), use vibratory polishing with a short-nap cloth and 1 μm alumina for 6 to 8 h. Final polish on a short-nap cloth with 0.05 μm alumina containing a saturated aqueous (cold) oxalic acid solution (!!! See section 4.4 in the Appendix.) for 1 to 2 h. If hydrogen peroxide is added, the polishing time is 1 h.
b. Polish with 6 μm diamond particle size on a hard synthetic cloth, then with 3 μm diamond particle size on a cotton cloth. Final polish with shock polishing using Ti EP1.
c. Polish electrolytically with Ti EP1, Ti EP2, or Ti EP3.

d. Polish chemically with Ti CP1 solution alternating with mechanical polishing and repeat this process several times.
Alpha Ti shows distinct color changes when examined in polarized light if the surface is free of deformation.

II. Ti-base alloys are used in the aircraft industry, as rotating components in turbine engines, and in high-performance engines.

Ti-Al with Sn, Zr, Mo, for example Ti-5Al-2Sn, Ti-6Al-4Zr-2Mo-2Sn, Ti-6Al-4V. Ti-Sn with Zr, Al, Mo, and Si.

Ti-2.5Cu (age hardenable).

Ti-base alloys consist of the alpha phase with a hexagonal close-packed crystal structure and with different proportions of the body-centered cubic beta phase. Quenched alloys have a martensitic (needlelike) alpha-beta microstructure. According to their heat treatment, they can be ductile to hard and are usually easy to prepare.

*Grinding:*

Wet grinding with SiC papers down to 1200 grit. Coat the 1200 grit paper with wax. The alloys require a careful and long grinding step on the fine papers.

*Polishing:*

a. Polish with 6 and 3 μm diamond particle size on a hard synthetic cloth and then on a cotton cloth. Final polish with alumina slurry to which a cold saturated aqueous oxalic acid has been added (!!! See section 4.4 in the Appendix.). If soft phases are present, add liquid soap to that mixture, or use $SiO_2$ suspension for the final polish.
b. Polish with 6 and 3 μm diamond particle size on a hard synthetic and then on a cotton cloth. Then shock polish with Ti EP1.
c. Polish electrolytically with Ti EP3.

III. Ti superalloys:

Ti-Al-V with a high content of Cr, Al, Mo, Zr, Fe, and Sn.

Superalloys are very hard and can only be ground and polished with automatic equipment.

*Grinding:*

Grind flat on diamond disks, then on 1200 grit SiC paper.

*Polishing:*

Polish with 6 and 1 μm diamond particle size on a hard silk cloth. Final polish for a few s on a short-nap velvet cloth with 0.05 μm alumina slurry.

## Electrochemical Polishing Recipes

### *Electrolytic Polishing*

| | | |
|---|---|---|
| Ti EP1 | 100 mL perchloric acid 60%<br>1000 mL methanol 99.8%<br>600 mL butyl cellosolve (wetting agent)<br>3.5 mL Vogel's reagent | Seconds to minutes<br>!!! See section 4.4 in the Appendix. |
| Ti EP2 | 200 mL perchloric acid 60%<br>700 mL ethanol 96%<br>100 mL 2-butoxy ethanol 99% | 15–60 s<br>30–65 V DC<br>!!! See section 4.4 in the Appendix. |
| Ti EP3 | 60 mL perchloric acid 60%<br>360 mL 2-butoxy ethanol 99%<br>600 mL methanol 99.8% | 45 s<br>58–66 V DC<br>!!! See section 4.4 in the Appendix. |

### *Chemical Polishing*

| | | |
|---|---|---|
| Ti CP1 | Use macroetchant Ti M1 | |
| Ti CP2 | Use microetchant Ti m1 | |
| Ti CP3 | 10 mL hydrofluoric acid 40%<br>10 mL nitric acid 65%<br>30 mL lactic acid 90% | Seconds to minutes<br>!! See section 4.4 in the Appendix. |

## Macroetching

| No. | Material | Etchant | Conditions |
|---|---|---|---|
| Ti M1 | Jodid (high-purity, iodide-process) Ti | 30 mL distilled water<br>10 mL hydrofluoric acid 40%<br>60 mL hydrogen peroxide 30% | Seconds to minutes<br>Swab until desired contrast is achieved.<br>!!! See section 4.4 in the Appendix. |
| Ti M2 | Ti alloys. Ti-Al-Mn alloys<br>Flow lines. Banded structure | 90 mL distilled water<br>10 mL hydrofluoric acid 40% | Up to 15 min<br>!!! See section 4.4 in the Appendix. |
| Ti M3 | Ti-Al-Mn alloys<br>Flow lines | 50 mL hydrofluoric acid 40%<br>50 mL glycerol 87% | Seconds to minutes<br>!!! See section 4.4 in the Appendix. |
| Ti M4 | Ti-Al-Mn, Ti-Al-V, and Ti-Al-Cr-Fe alloys<br>Flow lines | 95 mL distilled water<br>2.5 mL nitric acid 65%<br>1.5 mL hydrochloric acid 32%<br>0.5 mL hydrofluoric acid 40%<br>(Keller's reagent) | Seconds to minutes<br>!!! See section 4.4 in the Appendix. |

(continued)

## Macroetching (continued)

| No. | Material | Etchant | Conditions |
|---|---|---|---|
| Ti M5 | Pure Ti. Ti-base alloys<br>Ti-Al-Mo alloys | 50 mL distilled water<br>40 mL nitric acid 65%<br>10 mL hydrofluoric acid 40%<br>(HF content can be lowered) | 5–8 min<br>60–80 °C<br>!!! See section 4.4 in the Appendix. |
| Ti M6 | Welds | 200 mL distilled water<br>2 mL hydrofluoric acid 40%<br>10 g iron (III) nitrate<br>35 g oxalic acid | Seconds to minutes<br>50–60 °C<br>!!! See section 4.4 in the Appendix. |
| Ti M7 | Identification of alpha and beta Ti | 50 mL distilled water<br>50 mL hydrochloric acid 32% | Seconds to minutes |

## Microetching

| No. | Material | Etchant | Conditions |
|---|---|---|---|
| Ti m1 | Pure Ti and Ti alloys<br>Beta phase discolors blue.<br>Alpha and alpha prime phase discolor yellowish. | Heat tinting at 500–540 °C at regular furnace atmosphere | Cool on a metal substrate or a moist cloth. |
| Ti m2 | Pure Ti. Alpha Ti<br>Ti-Al-V-Sn alloys<br>Ferro titanium | 78 (20) mL distilled water<br>15 (5) mL hydrogen peroxide 30%<br>12 (10) mL aqueous potassium hydroxide 40%<br>(Concentration variable) | 3–20 s<br>Swabbing |
| Ti m3 | Pure Ti and Ti alloys<br>Grain-boundary etchant | 100 mL distilled water<br>5 mL hydrogen peroxide 30%<br>2 mL hydrofluoric acid 40% | 30–60 s<br>!!! See section 4.4 in the Appendix. |
| Ti m4 | For most Ti materials and Ti alloys<br>Ti-Mn and Ti-V-Cr-Al alloys | 85 (96) mL distilled water<br>10 (2) mL hydrofluoric acid 40%<br>5 (2) mL nitric acid 65%<br>(Concentration variable) | 3–20 s<br>!!! See section 4.4 in the Appendix. |
| Ti m5 | Ti-Al-Ni, Ti-Al-Sn, and Ti-Si alloys<br>Hydrides | 20 mL distilled water<br>45 mL glycerol 87%<br>25 mL nitric acid 65%<br>1 mL hydrofluoric acid 40% | 3–20 s<br>Swabbing<br>For Ti-Si alloys, use 3 mL hydrochloric acid 32%, instead of water.<br>!!! See section 4.4 in the Appendix.<br>!Explosive! |
| Ti m6 | Hydrides in Ti<br>Alpha-beta alloys | 30 mL lactic acid 90%<br>30 mL nitric acid 65%<br>1 mL hydrofluoric acid 40%<br>(Concentration variable) | 5–30 s<br>Do not store!<br>!!! See section 4.4 in the Appendix. |
| Ti m7 | For most Ti materials, especially for Ti-Al-V-(Sn) alloys | 100 mL distilled water<br>2–6 mL nitric acid 65%<br>1–3 mL hydrofluoric acid 40%<br>(Kroll's reagent) | 10–30 s<br>Swab 3–10 s<br>!!! See section 4.4 in the Appendix. |
| Ti m8 | Alpha Ti | 100 mL distilled water<br>1–10 mL hydrofluoric acid 40%<br>1–10 mL sulfuric acid 95–97% | Seconds to minutes<br>The etching time is reduced when glycerol is used instead of water.<br>!!! See section 4.4 in the Appendix. |

(continued)

## Microetching (continued)

| No. | Material | Etchant | Conditions |
|---|---|---|---|
| Ti m9 | Sn-containing Ti alloys | 100 mL distilled water<br>10 g sodium hydroxide<br>5 mL hydrogen peroxide 30% | Seconds to minutes |
| Ti m10 | Ti alloys | 190 mL distilled water<br>5 mL nitric acid 65%<br>3 mL hydrochloric acid 32%<br>1 mL hydrofluoric acid 40%<br>(Keller's reagent) | 1–20 s<br>Rinse in warm water.<br>!!! See section 4.4 in the Appendix. |
| Ti m11 | Ti alloys<br>Etches segregations | 100 mL distilled water<br>15 mL nitric acid 65%<br>10 mL hydrofluoric acid 40% | Seconds to minutes<br>Then immerse for 1 min into a 20 g/L solution of ammonium difluoride.<br>!!! See section 4.4 in the Appendix. |
| Ti m12 | Removal of Cu deposits on the surface | 100 mL water<br>25 mL nitric acid 65% | Seconds to minutes |
| Ti m13 | Removal of Sn deposits on the surface | 100 mL water<br>50 mL hydrogen peroxide 30%<br>2 mL nitric acid 65%<br>1 mL hydrofluoric acid 40% | Seconds to minutes<br>Rinse very well.<br>!!! See section 4.4 in the Appendix. |
| Ti m14 | Pure Ti and Ti alloys<br>Alpha and beta solid-solution phases are discolored.<br>Hydride and nitride needles, $O_2$-rich diffusion zones and beta phase remain unattacked.<br>Color etchant | 100 mL distilled water<br>50 mL ethanol 96%<br>2 g ammonium hydrogen fluoride<br>(Weck's reagent) | Seconds to minutes<br>Remove sample when a colored film changing from brown to blue appears.<br>Do not rinse after etching.<br>!!! See section 4.4 in the Appendix. |

## Electrolytic Etching

| No. | Material | Etchant | Conditions |
|---|---|---|---|
| Ti Em1 | Pure Ti | 25 mL distilled water<br>390 mL methanol 99.8%<br>350 mL ethylene glycol<br>35 mL perchloric acid 60% | 10–40 s at 5–10 °C<br>30–50 V DC<br>Stainless steel cathode<br>!!! See section 4.4 in the Appendix. |
| Ti Em2 | Pure Ti<br>Ti-base alloys | 80 mL glacial acetic acid<br>5 mL perchloric acid 60% | 1–5 min at 20–60 V DC<br>Stainless steel cathode<br>!!! See section 4.4 in the Appendix. |
| Ti Em3 | Ti and Ti-base alloys<br>Segregations<br>Color etchant | 35 mL distilled water<br>60 mL methanol 99.85<br>10 mL lactic acid 90%<br>5 mL phosphoric acid 85%<br>5 g citric acid<br>5 g oxalic acid | 10 s at 30–50 V DC<br>Stainless steel cathode<br>If necessary, dilute with equal amount of glycerol and etch for 60 s at 100–130 V DC.<br>!!! See section 4.4 in the Appendix. |

If a deposit forms during etching, immerse the sample immediately into a 10 to 30% aqueous solution of sulfuric acid.

**Physical Etching**

Since alpha titanium has a hexagonal crystal structure and is anisotropic, the microstructure can be examined with polarized light.

Y
La
Ce
Pr
Nd
Sm
Eu
Gd
Tb
Dy
Ho
Er
Tm
Yb
Lu

### *1.9.22 Yttrium, Lanthanum, and Rare Earth Metals*

**Preparation**

I. Pure yttrium, lanthanum, cerium, praseodymium, neodymium, samarium, europium, gadolinium, terbium, dysprosium, holmium, erbium, thulium, ytterbium, and lutetium.

Technical grade cerium-mischmetal such as Ce-La-Nd-Pr (CeMM).

The pure metals of these elements are soft and tend to corrode immediately when exposed to oxygen, water, and acids. Eu is nearly as reactive as Na and tarnishes instantly when exposed to air and is then very quickly destroyed by oxidation. La, Ce, Pr, and Nd are less reactive and can be worked for up to 30 min in air before oxide layers start forming. For longer times though, these metals, as well as Sm, Gd, and Yb, must be stored in an inert gas atmosphere. Only the rare earth metals from Tb to Lu and Y remain unaffected by air exposure. Because of their chemical reactivity and their low strength, pure rare earth metals have no technical application.

*Sectioning:*

a. Cut Eu and Yb with a knife or a razor blade (Eu under inert atmosphere).
b. All rare earth metals except for Eu can be saw cut or sectioned with a SiC or copper-diamond cutoff wheel. Low pressure and ample cooling are necessary. Use a water-based coolant, but rinse and dry the sample immediately with ethanol and acetone. Do not use volatile and combustible hydrocarbons, such as chlore or fluor hydrocarbons, as a coolant. (!!! See section 4.4 in the Appendix.).
La and Ce chips are highly ignitable (!!!), Pr, Nd, and Sm chips react with chlorinated or hydrofluorocarbons. Treat the chips immediately with 5 to 10% hydrochloric acid in alcohol for safe disposal.
Oxidized surfaces must first be removed by filing. It should be noted, that oxidation occurs preferably along the grain boundaries and may be several millimeters deep.

*Grinding:*

Metallographic samples of rare earth metals can be ground on SiC papers from 320 to 2400 grit, using only low pressure. Use kerosene or petroleum as a coolant. The finer papers should be coated with wax and a faster wheel speed is recommended.

*Polishing:*

a. After grinding to 2400 grit on SiC paper, polish the sample on a cotton cloth and a short-nap velvet cloth with a 1 and 0.25 μm diamond particle size gel (water free) and a water-free lubricant. Low pressure and long polishing times are required.
b. The less reactive group of rare earth metals can be polished on a felt cloth with a 1 μm water-based alumina slurry. Polish until a uniform and transparent oxide film has formed.
c. There is no suitable polishing method for Eu. Its surface can only be removed by ultramilling or with a diamond blade.
d. La and Ce can be polished electrochemically and anodized with RE EP1 and Yb with RE EP2.
e. Chemical polish (slurry polish) with RE WP1 can be used for Y, Gd-Tm, and Lu, but not for Yb and La-Nd. Use RE WP2 for La-Gd and RE WP3 for Yb.
f. Etch polish all rare earth metals except Eu with RE m9.

II. Sm-Co alloys and Fe-Nd-B alloys are used for permanent magnets.

Mg-Ce-Zr alloys are used for parts in the aviation industry.

CeMM-Mg-Fe alloys are used for lighters.

Only small additions of rare earth metals are used as alloying elements. They form intermetallic phases of special properties, e.g., exceptional magnetic properties. However, they are highly prone to corrosion; they are brittle and hard. Other phases such as oxides, carbides, and halides interfere with the sample preparation.

Microscopic examination should be carried out immediately after the preparation, since an oxide film is formed immediately and some phases are decomposing in air. Therefore, the samples should be stored in an inert gas atmosphere or submersed in petroleum.

*Grinding:*

Grind on SiC papers with a water-free coolant. Coat all papers from 1000 grit and finer with wax. Ce-Fe-Mg and Fe-Nd-B alloys may contain elemental rare earth metals that are sensitive to water. Grinding can also be done dry, but papers have to be changed frequently. Grind very carefully and examine whether secondary phases are being retained.

*Polishing:*

Use 3 and 1 μm diamond particle size gel and a water-free coolant on a cotton cloth and a short-nap velvet cloth.

## Electrochemical Polishing Recipes

(RE = abbreviation for rare earth)

*Electrochemical Polishing and Anodizing*

| | | | |
|---|---|---|---|
| RE EP1 | 100 mL<br>1–2 g | distilled water<br>potassium hydroxide | Seconds to minutes. 22 V DC<br>Rinse in distilled water, then with methanol.<br>Use for La and Ce. |
| RE EP2 | 100 mL<br>10 mL | methanol 99.8%<br>hydrochloric acid 32% | Seconds to minutes<br>30 V DC<br>Rinse with methanol. |

*Chemical-Mechanical Polishing (Etch Polishing)*

| | | | |
|---|---|---|---|
| RE WP1 | (a) 20 mL<br>15 mL<br>10 mL<br>5 mL<br>1 mL | lactic acid 90%<br>nitric acid 65%<br>glacial acetic acid<br>phosphoric acid 85%<br>sulfuric acid 95–97% | 10–15 s with (a)<br>Rinse with distilled water.<br>For Gd-Lu alloys and Y |
| | (b) 25 mL | dimethyl formamid | 10–15 s with (a) and (b). Rinse with methanol.<br>For La-Nd alloys.<br>!!! See section 4.4 in the Appendix. |
| RE WP2 | 25 mL<br>25 mL<br>25 mL<br>25 mL | dimethyl formamid<br>lactic acid 90%<br>glacial acetic acid<br>nitric acid 65% | 10–15 s<br>Rinse with ethanol.<br>!!! See section 4.4 in the Appendix. |
| RE WP3 | 58 mL<br>22 mL<br>5–8 mL | phosphoric acid 85%<br>methanol 99.8%<br>nitric acid 65% | 10 s<br>Rinse with distilled water, then with methanol. |

## Microetching

| No. | Material | Etchant | Conditions |
|---|---|---|---|
| RE m1 | For most alloys of RE metals<br>CeMM alloys | Exposure to air | min to h<br>From RT to 200 °C |
| RE m2 | Pure Gd. For most RE metals and their alloys. Sm-Co alloys | 75 mL glacial acetic acid<br>25 mL hydrogen peroxide 30% | 5–15 s |
| RE m3 | Pure Gd. RE-Co alloys<br>Grain-boundary etchant | 45 mL ethanol 96%<br>1 mL nitric acid 65%<br>(Nital etchant) | 1–3 min |
| RE m4 | Dy-, Er-, Gd-, Ho-, and La-base alloys<br>RE-Co alloys | 20 mL lactic acid 90%<br>15 mL nitric acid 65%<br>10 mL glacial acetic acid<br>5 mL phosphoric acid 85%<br>1 mL sulfuric acid 95–97%<br>(Concentration variable) | 10–15 s |
| RE m5 | Sm-Co and Sm-Fe-Co-Cu alloys | 100 mL distilled water<br>10 g ammonium persulfate | 5 s<br>Swabbing |
| RE m6 | Sm-Co and Sm-Fe-Co-Cu alloys<br>Color etchant | 100 mL distilled water<br>10 g potassium disulfate<br>Few drops of hydrochloric acid 32% | Seconds to minutes |
| RE m7 | $CeC_2$<br>For polarized light examination | 30 mL distilled water<br>30 mL glacial acetic acid<br>30 mL nitric acid 65% | Seconds to minutes |
| RE m8 | Sm-Co alloys<br>Grain-boundary etchant | 100 mL distilled water<br>8 g chromium (VI) oxide<br>2 g sodium sulfate | 5–10 s<br>!!! See section 4.4 in the Appendix. |
| RE m9 | Sm-Co alloys | 100 mL distilled water<br>1 mL glacial acetic acid<br>1 mL nitric acid 65% | Few s |
| RE m10 | RE-Al alloys | 955 mL distilled water<br>25 mL nitric acid 65%<br>10 mL hydrochloric acid 32%<br>5 mL hydrofluoric acid 40%<br>(Keller's reagent) | Few s<br>!!! See section 4.4 in the Appendix. |
| RE m11 | RE-Mg alloys | 25 mL water<br>75 mL diethylene glycol<br>1 mL nitric acid 65% | Few s |
| RE m12 | Nd-Fe-B alloys<br>Phase identification | 100 mL ethanol 96%<br>2 mL ammonium hydroxide solution 25% | Few s |

## Physical Etching

Rare earth metals and their alloys are anisotropic except for Ce, Eu, and Yb; therefore their phases can be easily identified with polarized light, using the magneto-optical Kerr effect. In order to enhance the Kerr effect, the surface should be coated with ZnSe or $Pt/O_2$.

When using the Bitter technique, magnetic domains can be revealed. This method is used for permanent magnets and Sm-Co alloys, especially $SmCo_5$. Description of this method can be found in section 4.1 in the Appendix.

Zn

### *1.9.23 Zinc*

**Preparation**

I. Pure zinc (99.95 to 99.90) is used for pressure die casting.

Commercial zinc (99.5 to 97.5) contains Pb, Cd, Sn, and Fe.

Zn is used for galvanic coatings on iron and steel against corrosion attack; it is also used for containers and as cathodes in batteries.

Zn tends to form deformation and smear layers, ghost structures, and deformation twins during sample preparation. Mechanical pressure and heat generation should therefore be avoided.

*Grinding:*

a. Use microtoming with a tungsten carbide or diamond knife for pure Zn, and grinding and polishing can then be omitted.
b. Grind on a stationary grinder using wax or stearin-coated SiC to 1200 grit. Since the deformation layer of the individual grinding steps will extend far deeper than the grinding marks on the surface, it is necessary to grind longer on the finer grit papers and also by changing the grinding direction.

*Rough Polishing:*

Polish with 1 μm alumina on a felt cloth treated with soap to remove the deformation layer from the grinding steps.

*Polishing:*

a. With 1 μm alumina slurry on a short-nap cloth, or with 6, 3 and possibly 0.25 μm diamond particle size. Then polish on a cotton cloth with a suspension of:
   120 mL warm distilled water
   20 mL aqueous ammonium tartrate solution 5%
   1 g magnesia
   This suspension must be filtered through a fine silk or cotton cloth.
b. Etch polishing:
   Polish with 6 μm diamond particle size for 4 min, then etch for 3.5 min.
   Polish with 3 μm diamond particle size for 6 min, then etch for 1.5 min.

Polish with 1 μm diamond particle size for 8 min, then etch for 30 s.
Polish with 0.25 μm diamond particle size for 10 min, then etch for 10 s.
Polish with 0.05 μm alumina particle size and etch for 5 s.
Suitable etchants are Zn m4 or Zn m5.

c. Polish electrolytically with Zn EP1 and Zn EP2.
d. Polish chemically with Zn CP1 and Zn CP2.

II. Zn pressure die-casting alloys such as Zn-Al with Cu and Mg are used for household and kitchenware and for small engine parts.

Zn sheet and plate alloys with Pb, Cd, Cu, Ti separately or combined are used for parts in the automobile industry and for home and office furniture parts.

Hard solders: Zn-6Sn-4Cu.

Follow the sample preparation as for I, but use rotating grinding wheels and shorter times. To prevent relief formation of the harder constituents within the soft Zn matrix, it is recommended to use a short-nap velvet cloth for the final polishing step.

Zn, Cu, and Pb cannot be polished on the same cloth.

After the final polishing is completed, the sample should be cleaned with alcohol, then immersed into distilled water. The wet sample should be etched immediately.

## Electrochemical Polishing Recipes

*Electrolytic Polishing*

| | | | |
|---|---|---|---|
| Zn EP1 | 100 mL | distilled water | 40–50 s |
| | 20 g | chromium (VI) oxide | 60 V DC |
| | | | Pt or Zn cathode |
| | | | !!! See section 4.4 in the Appendix. |

| | | | |
|---|---|---|---|
| Zn EP2 | 50 mL | phosphoric acid 85% | 1 h or longer. 4–6 V DC |
| | 50 mL | ethanol 96% | Stainless steel cathode |
| | | | !!! See section 4.4 in the Appendix. |

*Chemical Polishing*

| | | | |
|---|---|---|---|
| Zn CP1 | 100 mL | distilled water | Up to 30 min |
| | 5 mL | nitric acid 65% | The reaction layer that formed is water soluble. |
| | 20 g | chromium (VI) oxide | |
| | 1.5 g | sodium sulfate | !!! See section 4.4 in the Appendix. |

(continued)

*Chemical Polishing (continued)*

| | | | |
|---|---|---|---|
| Zn CP2 | 1000 mL | distilled water | 5–10 min |
| | 200 g | sodium dichromate | |
| | 6–9 mL | sulfuric acid 95–97% | |

## Macroetching

| No. | Material | Etchant | Conditions |
|---|---|---|---|
| Zn M1 | Pure Zn<br>Zn alloys without Cu<br>Foundry alloys | 50 mL distilled water<br>50 mL hydrochloric acid 32%<br>or: only hydrochloric acid 32%<br>or: only nitric acid 65% | 15 s<br>Remove deposit under running water. |
| Zn M2 | Zn alloys (with Cu) | 100 mL distilled water<br>20 g chromium (VI) oxide<br>1.5 g sodium sulfate **(anhydrous)**<br>(Palmerton's reagent) | Seconds to minutes<br>If using crystalline sodium sulfate (**$Na_2SO_4.10H_2O$**) increase the addition to 3.5 g<br>!!! See section 4.4 in the Appendix. |
| Zn M3 | Zn-base alloys | 95 mL ethanol 96%<br>5 mL hydrochloric acid 32% | Seconds to minutes |

## Microetching

| No. | Material | Etchant | Conditions |
|---|---|---|---|
| Zn m1 | Pure Zn and low-alloy Zn<br>Color etchant | 100 mL Klemm's stock solution<br>2 g potassium disulfite<br>(Klemm's reagent I)<br>**Klemm's stock solution:**<br>300 mL distilled water (40 °C)<br>1000 mL sodium thiosulfate | 30 s |
| Zn m2 | Pure Zn (technical grade)<br>Zn-Co and Zn-Cu alloys | 100 mL distilled water<br>10 g sodium hydroxide | 1–5 s |
| Zn m3 | Pure Zn. Zn-Cu-Al alloys | 100 mL distilled water or ethanol 96%<br>1–5 mL hydrochloric acid 32% | Seconds to minutes |
| Zn m4 | For most Zn materials and Zn alloys<br>Especially for Pb-containing sheet zinc<br>Zn-Cu alloys | 200 mL distilled water<br>20 g chromium (VI) oxide<br>1.5 g sodium sulfate<br>(Palmerton's reagent) | 2–3 min<br>Reduce the sodium sulfate content to half for Cu-containing alloys. Rinse in 20% aqueous chromium (VI) oxide solution.<br>!!! See section 4.4 in the Appendix. |
| Zn m5 | All Zn alloys<br>For pressure die castings<br>Zn coatings | 100 mL distilled water<br>5 g chromium (VI) oxide<br>0.5 g sodium sulfate<br>(modified Palmerton's reagent) | 2–3 min<br>Rinse in 20% aqueous chromium (VI) oxide solution.<br>!!! See section 4.4 in the Appendix. |
| Zn m6 | Zn-Al and Zn-Cr alloys<br>Zn-Fe layers on galvanized steel or ferrous materials<br>Grain-boundary etchant | 100 mL distilled water<br>1 mL nitric acid 65% | Seconds to minutes<br>Don't keep solution longer than 1 h!<br>To avoid staining, rinse with 20% aqueous chromium (VI) oxide solution.<br>!!! See section 4.4 in the Appendix. |

(continued)

## Microetching (continued)

| No. | Material | Etchant | Conditions |
|---|---|---|---|
| Zn m7 | Zn alloys containing metals of the Fe group<br>Zn galvanized coatings | 70 mL distilled water<br>30 mL ethanol 96%<br>0.3 g picric acid | s to 3 min<br>!!! See section 4.4 in the Appendix. |
| Zn m8 | Zn alloys<br>Banded structures in extrusion alloys | 100 mL distilled water<br>11 g ammonium persulfate<br>1 g citric acid | Seconds to minutes |
| Zn m9 | Zn alloys containing noble elements, e.g. Ag, Au, Cu | 1250 mL distilled water<br>30 mL hydrochloric acid 32%<br>4 g iron (III) chloride | Seconds to minutes |
| Zn m10 | Zn-Cu, Zn-Fe, Zn-Mg, Zn-Ni, and Zn-Pb alloys<br>Zn-rich phase discolors dark. | 900 mL distilled water<br>60 mL saturated aqueous copper (II) nitrate solution<br>122 g potassium hydroxide<br>75 g potassium cyanide<br>6.5 g citric acid<br>(Schramm's reagent)<br>(Concentration variable) | Seconds to minutes<br>Mixing sequence must be followed, otherwise hydrocyanic acid may develop.<br>Care must be taken when disposing of the etchant!<br>!!! See section 4.4 in the Appendix. |

## Electrolytic Etching

| No. | Material | Etchant | Conditions |
|---|---|---|---|
| Zn Em1 | Pure Zn and low-alloy Zn | 100 mL distilled water<br>25 g sodium hydroxide | 15 min<br>6 V DC<br>Cu cathode |
| Zn Em2 | Zn-Cu alloys<br>Distinction between thegamma and epsilon phases | 100 mL distilled water<br>10 g chromium (VI) oxide | Seconds to minutes<br>12 V DC<br>Pt cathode<br>!!! See section 4.4 in the Appendix. |

## Physical Etching

Zinc can be examined in polarized light because of its hexagonal crystal structure.

# 2 Ceramography

Ceramic materials are finding increased applications in many areas of technology, and the analysis of ceramic materials is therefore becoming more common in materials testing laboratories. Among others, light and electron microscopy play an important role. Like metallic materials, the prepared specimens are preferably examined with reflected light or scanning electron microscopy. The previously described methods for metallographic sample preparation can, with some modifications, also be used for ceramic materials (see Chapter 1 "Metallography"). The microstructural examination of ceramic materials is known as "ceramography."

The following methods for ceramographic sample preparation apply for all ceramic materials and are similar to metallic materials in their variety. Since the traditional ceramics, such as porcelain, stoneware, glasses, and fireproof materials, contain mostly a high proportion of silicates, they are often examined as thin sections using transmitted light. Ceramography has been particularly successful for the examination of advanced ceramics. *Advanced ceramics* are inorganic, nonmetallic materials with a *precisely defined microstructure.* While the traditional ceramics are made from natural materials such as clays, sands, or kaolins, the base for advanced ceramics are pure oxides, nitrides, carbides, borides, and their mixtures with a definite composition, particle shape, and particle size distribution. Usually they are powders that are compressed and sintered into compacts. The microstructure is strictly optimized (tailored) because properties of the compacted powder are even more closely related to the microstructure (porosity, pore size, and grain size distribution) than in metallic materials.

## 2.1 Sample Preparation for Ceramic Materials

As in metallographic sample preparation, the preparation of ceramic samples requires the same steps: sampling, mounting, identifying, grinding and polishing, cleaning, and etching. In principle, the descriptions in Chapter 1 are also applicable for ceramic materials. However, the brittleness, the often very high hardness, the porosity, and the chemical resistance of ceramic materials present specific difficulties, which are taken into account in the methods and recipes described. Similar to the preparation procedures for metal or polymer samples, the preparation methods for ceramics are also varied.

> In any case, the prepared specimen must clearly reveal the microstructure for macroscopic and microscopic examination and must not contain any artifacts that were generated by the sample preparation.

This applies also to metal ceramics (*cermets*) and *layer composite materials,* which are also described in this chapter. These types of materials present further difficulties, such as large differences in hardness and in chemical behavior between the metallic and ceramic constituents.

## 2.2 Sampling

Sampling is done almost exclusively by sectioning with motorized diamond cutoff machines using water or oil emulsions as lubricant and coolant (see section 1.2 in Chapter 1). Alcohol or low-viscosity mineral oils can be used for sectioning and grinding of materials that are affected by water (beta $Al_2O_3$, CaO, MgO). Ceramics can be easily sectioned at high or low speeds. Low-speed sectioning machines (1000 rpm max) are suited for small and intricate samples; they cause less in-depth damage by breakouts than the high-speed sectioning machines (1000 to 5000 rpm) used for larger ceramic parts. When selecting cutting disks, it is important to know not only the type of ceramic to be cut, but also the disk substrate and the thickness of the disk. Typically, thin (0.3 to 0.5 μm) plastic-bonded diamond disks, also called "soft" disks, are used.

## 2.3 Mounting

Continuing from the description in section 1.3 (in Chapter 1), it should be noted that for automatic sample preparation the specimens can be glued directly onto a sample holder or can be clamped together or mounted. The mounting material should be either very hard or have good abrasion resistance. Adding a hard material to the mounting material is only necessary when good edge retention of the sample is required.

> Before mounting, the etching method to be used later should already be known. For example, samples that will be thermally etched or etched in a

molten salt bath should be mounted such that the sample can be easily removed from the mount before etching; when etching in boiling solutions, the mounting material must be chemically resistant.

*Infiltration* (see section 1.3.3 in Chapter 1) is a very important technique used in ceramography, and it is the only method that guarantees exact reproducibility of the actual pore size and distribution. Therefore, infiltrating any material that has open or closed porosity with a low-viscosity resin is recommended to prevent misinterpretation caused by breakouts or smearing of the pores. The resin penetrates the interconnected pores, creating a solid bond between specimen and infiltration material. Specimens with open porosity should be infiltrated before sectioning, and those with a porosity content of 5 to 14% can be infiltrated before grinding. The quality of the infiltration should be re-examined after the fine-grinding step. If the adhesion between specimen and infiltration material is poor, the specimen should be re-infiltrated; the specimen may remain in the sample holder. Porosity levels of 5% and less do not require infiltration and at that level is also hardly successful. Specimens that will be thermally etched cannot be infiltrated; the infiltration material decomposes with a volume expansion at elevated temperatures and may cause the sample to "explode." If infiltration is absolutely necessary, it then has to be very carefully burnt before thermal etching; this may generate poisonous fumes (!!! see section 4.4 in the Appendix).

## 2.4 Identification

For the identification of ceramic samples, except for stamping, refer to the methods described in section 1.4 (in Chapter 1).

## 2.5 Grinding and Polishing

The methods described for metallic specimens in section 1.5 (in Chapter 1) for preparing the specimen surface can generally also be applied for ceramic and metal-ceramic specimens. It is important, however, to always know the inherent material characteristics, especially its brittleness.

### *2.5.1 Mechanical Grinding and Polishing*

As in metallography, mechanical grinding and polishing methods are also most common in ceramography, where the material is successively removed from the surface by the grinding and polishing media.

Because of their high brittleness, the ceramic specimens show very little deformation and therefore contain practically no smear layers such as occur with metals (see Fig. 1.9 in Chapter 1). However, breakouts of grains or grain fragments, as well as surface cracks that are generated by sectioning, occur frequently. In addition, material removal occurs by cutting action of the grinding and polishing material and to a lesser extent by the adhesion effect with the polishing substrate.

The structure adjacent to the surface of a mechanically ground and polished ceramic specimen is shown in the schematic in Fig. 2.1. This can lead to misinterpretation due to damaged microstructures, as shown in the schematic in Fig. 2.2(a) and (b). Despite their very low deformability, deformation twins can in some instances form during grinding; dislocations that are mostly parallel to the grinding and polishing scratches can also be observed.

*Coarse (flat) grinding* (60 μm grit disks) and *fine grinding* (20 μm grit disks) are preferably done on plastic-bonded diamond disks with running water. Diamond grinding disks show excellent cutting consistency with long durability. They are suitable for all advanced ceramics. Ceramics with a Vickers hardness lower than 1000 HV can be ground on steel net cloth or hard perforated synthetic cloth with diamond grits of 40 and 25 μm and with an alcohol-based lubricant. On the other hand, silicate ceramics and refractories are polished on SiC papers.

*Automatic grinding* provides a choice of rotating speed and pressure and their reproducible adjustments. Even though a higher surface removal rate is achieved with increasing rotational speed and pressure, it may cause more extensive sur-

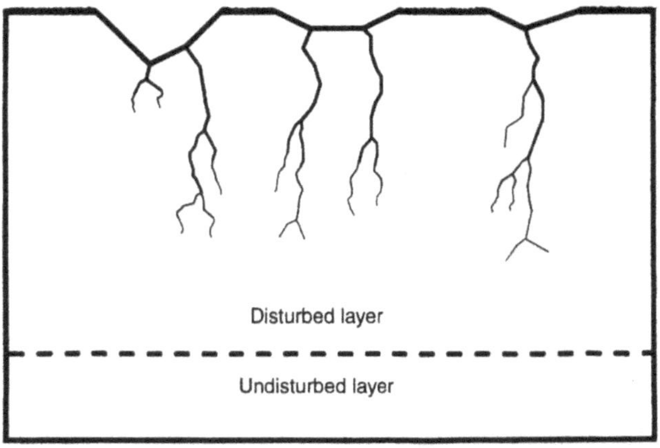


**Fig. 2.1** Surface of a mechanically sectioned and rough polished sample (schematic cross section vertical to the specimen surface)

face damage. High speeds and high pressures are recommended only for coarse grinding, and both parameters should be reduced for fine grinding (e.g., to 150 rpm, 90 N) in order to reduce surface damage and breakouts. This is particularly important for porous materials. Running water is usually sufficient as a *coolant (lubricant)* during grinding; any exceptions are noted in the recipes.

The surface damage generated during grinding has to be removed during lapping and polishing. *Lapping* has proven to be a very effective method (see section 1.5.1 in Chapter 1). The specimen surface remains flat, and no further damage is introduced. A suitable lapping disk consists of a composite with a matrix of plastics containing metal particles of various hardness. The lapping medium contains diamond grit particles in a lubricant selected for the particular system. The choice of pressure, speed, and time depend on the hardness, brittleness, and porosity of the material. Lapping disks of different hardnesses combined with diamond grits of various sizes can be selected. Several lapping steps can be included between the grinding and polishing steps. The following preparation methods all include a lapping step using 6 μm diamond particle size.

After the lapping step, *polishing* is performed with 3 μm diamond particle size; the surface should have a fine-ground appearance after lapping when examined with the microscope; a 6 μm diamond particle size polish is necessary if the specimen surface still appears rough. A hard synthetic cloth impregnated with an

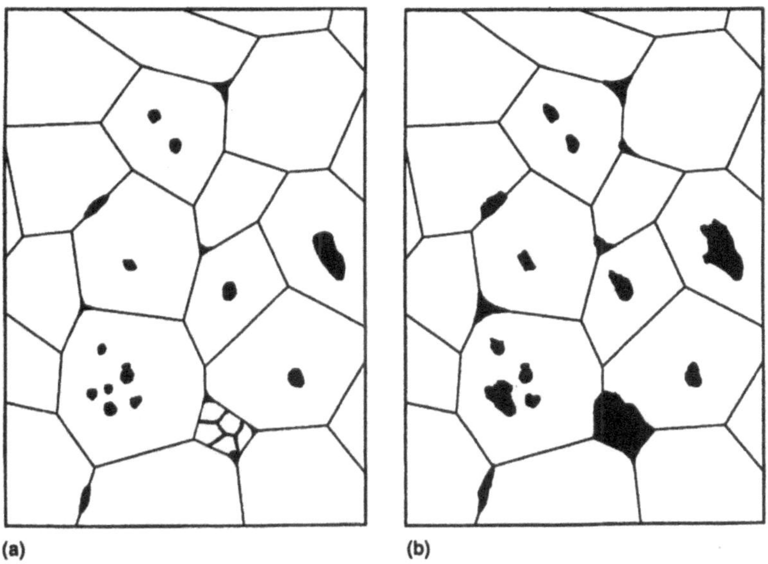


**Fig. 2.2** Schematic of (a) the true microstructure and (b) a microstructure distorted by breakouts

alcohol-based lubricant is used as a polishing substrate. The specimen is polished until no further changes on the surface can be microscopically observed. The next steps are to polish with smaller particle sizes, 3 and 1 μm diamond particle size, on a hard silk or cotton cloth with less pressure. All breakouts and scratches are to be removed during these steps. The polishing progress can be monitored by microscopic observation. If desired, a slight grain-boundary relief can be produced with an alumina polish on a velvet cloth. *Metal ceramics* should be polished with short times and high pressure on hard cloths to prevent relief formation at the metal/ceramic interface. A final polish with an alkaline $SiO_2$ suspension on a chemically resistant velvet cloth at a short polishing time is recommended; this produces scratch-free surfaces with well-defined metal/ceramic interfaces.

The recommendations in the recipes regarding the substrates used for grinding, lapping, and polishing, as well as sample pressure, time, and rotational speed can be optimized and adjusted to individually different materials.

### 2.5.2 Other Grinding and Polishing Methods

The other mechanical grinding and polishing methods for metallography referred to in section 1.5 (in Chapter 1) have little application in ceramography. However, *microtome cutting* and *ultramilling* (see section 1.5.2 in Chapter 1) may in some cases be very useful for the preparation of metal-ceramic composites used in the electronics industry (microcircuits, chips, condensers), for biomaterials (heart pacemakers, dental implants), or for coatings (ceramic protective or insulating coatings). *Electrolytic grinding and polishing methods* (see section 1.5.3 in Chapter 1) can only be used for materials that are electrically conductive. This applies for some carbides and metal ceramics, but the present experience is limited and not very good. The same applies for *chemical methods* and *combined polishing methods* (see sections 1.5.4 and 1.5.5, respectively, in Chapter 1). They are limited because of the highly reactive chemicals or high temperatures needed, making the handling very difficult.

## 2.6 Cleaning

The cleaning of metallographic specimens as described in section 1.6 (in Chapter 1) applies in every detail for the ceramographic specimens. Breakouts and porosity are quite common in ceramic specimens, and they must be specifically addressed. It is absolutely necessary to remove all abrasion from the pores and breakouts because the abraded particles are hard and can damage the surface. It is also very important to ascertain that all entrapped and remaining liquid within the pores and breakouts has been completely evaporated. All the cleaning methods as described in Fig. 1.14 can be used.

## 2.7 Ceramographic Etching

It is good practice to examine a ceramic specimen in its as-polished condition under the microscope before etching. This permits the evaluation of the amount of pores and their distribution and possibly the evidence of breakouts. Inclusions, contaminations, and cracks should be analyzed first on the unetched specimens. They will have to be etched, though, to reveal the grain boundaries, phases, and other microstructural features.

Etching in ceramography is nearly as diverse as in metallography (see section 1.7 in Chapter 1). The ceramographic etching methods are listed in Fig. 2.3.

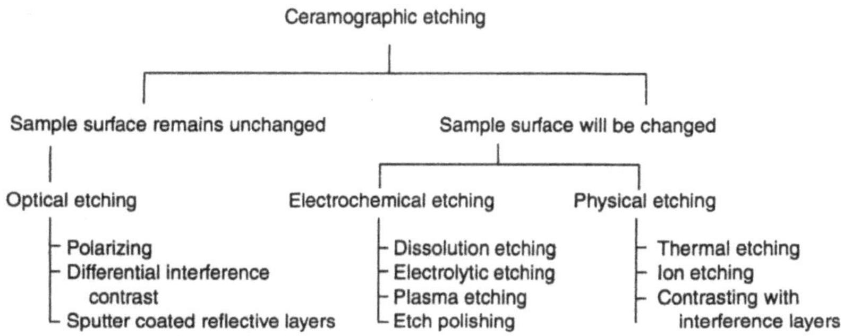


**Fig. 2.3** Etching techniques used in ceramography

### *2.7.1 Optical Etching*

Optical contrasting techniques (see section 1.7.1 in Chapter 1), such as darkfield, polarized light, and phase contrast play only a minor role in the examination of microstructures in advanced ceramics prior to etching. They are more successfully applied in combination with other etching methods (see sections 2.7.2 and 2.7.3 in this chapter). This is related to the fact that chemically or thermally etched ceramic specimens, examined under the microscope with reflective brightfield illumination, produce a low contrast and whitish image. Ceramics permit the light ray to penetrate the surface, resulting in scattering and internal reflections of the light, rendering it difficult to obtain a true image of the surface.

To eliminate light scattering and to improve the reflectivity, it is recommended that the specimen surface be coated with a *reflective layer*. Such a layer should be between 5 and 10 nm thick and can be sputter coated with either Au or Al. This way, the details of the structure's topography are truly revealed; however, the specific color differences of the individual phases are not developed.

Ceramics are often optically anisotropic. However, because of their low reflectivity, *polarization* for qualitative evaluation of specimens can only be used under certain favorable conditions. When examining thin sections, though, observation with polarized light is of increased importance since information such as differences in phases and stresses, as well as inclusions, can be obtained. Further details for the preparation of thin sections can be found in the article by R. Telle and G. Petzow: Light Microscopy—Sample Preparation Techniques, in *Materials Science and Technology* (Volume 2A, *Characterization of Materials,* Part I, Volume editor: E. Lifshin, edited by: R.W. Cahn, P. Haasen, and E.J. Kramer, VCH Verlagsgruppe, Weinheim, Germany, 1992, p 413–419). However, the use of a polarizing filter when working with incident light is still recommended since *polarizing* the light beam decreases light scattering and provides more image contrast.

*Differential interference contrast* is another technique used to reveal the surface details. The microstructure appears three-dimensional, with grain boundaries and phases in relief. However, coating the surface prior to examination with a thin reflective layer is recommended.

Ceramics with good reflectivity ($B_4C$, SiC, and transition-metal carbides and borides) can easily be examined with *oil-immersion objective lenses,* like any metallic material. However, if the ceramic is light transmitting, the effect of internal reflections and the light scattering effect in the specimen becomes excessive, and the use of oil immersion objective lenses is no longer possible.

*Darkfield illumination* is rarely used in ceramography. If it is difficult to differentiate between a pore and a phase, darkfield illumination may be useful for the identification. Darkfield illumination produces a bright halo around a large pore, while different phases remain dark. Copper-containing materials, such as oxide superconductors (e.g., Y-Ba-Cu-O-alloys) may contain cupric oxide ($Cu_2O$), which appears in its typical purple color when viewed in darkfield illumination.

## 2.7.2 Electrochemical (Chemical) Etching

The generally good chemical resistance, even at elevated temperatures, is a favorable property of ceramic materials; naturally this becomes a disadvantage for the typical chemical etching (*dissolution etching*). Very aggressive chemical reagents are usually necessary; they are often concentrated boiling solutions or molten salts. Protection and safety measures *must always* be well observed (!!! see section 4.4 in the Appendix).

Single-phase ceramics can normally be etched without problems. Grain boundaries are preferentially attacked, but grain contrast can be etched at different rates depending on its orientation. Multiphase ceramics and metal ceramics are much more difficult to etch. Their individual phases often have a very different chemical potential that may result in the dissolution of one phase. Differences in the crystal structure of one and the same compound (e.g., modifications of SiC: cubic/hexagonal) cannot be distinguished by chemical etching.

> The following undesirable effects can occur with chemical etching: nonuniform etching of grain boundaries, removal of very small grains during etching, dissolution of grain-boundary phases, exposure of scratches, artifact layers disguising the true microstructure, remnants of chemicals used for etching, enlarged pores and cavities, and poor reproducibility.

When etching with molten salts, tube, muffle, or box furnaces that open at one side and provide easy access for the small crucibles are preferred. Vertical tube furnaces with suspended or upright crucibles are more difficult to work with. The crucible material must be resistant to the melt. The recommended etching times are only approximate. They have to be carefully established for each individual material since the material can easily be overetched.

Ceramics have a poor electrical conductivity, and *electrolytic etching* is only possible for few materials. The success of such an etch depends on a variety of material factors, such as grain size, density, content of glass phases, and the degree of purity; the recommendations are often not effective and have to be modified accordingly. When using literature recommendations, it is important to review the exact composition of the sample material; even minute differences may result in major differences of the etching behavior.

*Plasma etching* has to date only been successfully used for silicon nitride ($Si_3N_4$)-base materials and for liquid-phase-sintered SiC. It is carried out in a commercially available plasma etching apparatus. This apparatus consists of a radiofrequency generator and a reaction chamber. Two inlets provide a gas mixture of $O_2$ and $CF_4$. The high-frequency electromagnetic waves produce fluor radicals that react with the $Si_3N_4$ matrix and form gaseous $SiF_4$. Nitrogen and carbon dioxide are emitted as reaction gases. Contrary to dissolution etching, the grain-boundary phases remain unattacked by plasma etching, and this permits the identification of pores, their size, and shape. The intensity can be controlled by changing the partial pressure of the oxygen.

*Etch polishing* can produce a chemical attack of grains and grain boundaries in some ceramics by adding a chemical to the polishing suspension. The glass phase in $Al_2O_3$ can be etched by adding NaOH or a modified Murakami solution.

Grain-boundary phases in AlN can be revealed by adding NaOH to the polishing suspension or using an alkaline final polishing suspension of $SiO_2$. When using semiautomatic equipment, the polishing wheel can be modified and turned into a container where the etching suspension is redirected to the wheel center with a special attachment.

## 2.7.3 Physical Etching

*Ion etching* does not always successfully produce clear images and has limitations when used alone. Satisfactory results were obtained only for porcelain, Si/SiC, and $ZrO_2$ ceramics. Ion etching helps to establish depth concentration profiles that can be measured with additional sensitive analytical equipment; it can also be used for specimen thinning for transmission electron microscopy.

*Thermal etching* is an important method used to reveal the ceramic microstructure. The polished and "de-mounted" samples are brought to red heat in a furnace; the temperature and time are selected so that the heat treatment does not alter the microstructure.

When using this etching method, it is important to know the material behavior during thermal etching, since the etching temperature depends on the previous heat treatment of the material. It should be sufficiently high to ascertain the grain boundary and grain surface diffusion necessary for the etching process. Suitable etching temperatures for most materials are approximately 100 to 250 °C below the sintering temperature. Too high temperatures or too long etching times may result in grain growth by volume diffusion. Precipitation and transformation reactions can also occur in some materials. Proportions of glass phases can melt, and grain-boundary phases can evaporate. Oxide ceramics are etched under air or hydrogen atmosphere, carbide ceramics in vacuum or noble gas, nitride ceramics in a nitrogen atmosphere. Thermal etching is usually carried out under atmospheric pressure.

The deposition of *interference layers* for phase contrasting in ceramics does not show as good results as it does in metals. Ceramics have a much lower reflectivity coefficient than metals and are partially light transmitting. The conditions for multiple reflectivity and the interface of coating/ceramic, which produce interference, are not present. Ceramic microstructures can only be satisfactorily contrasted with interference layers under the following conditions:

- The microstructure should be multiphase, and the phases should be opaque. Reactive sputtering coating materials such as Pt, Fe, or Pb cathodes are suitable in this case. Approved materials for vapor deposition are ZnSe, ZnTe, and ZnS.

- If the microstructure is multiphase and the phases are light transmitting, a coating material with a low refractive index should be chosen to increase the multiple reflectivity. Coating materials such as cryolith ($Na_3AlF_6$) are suitable.
- Layers and composite systems of metal and ceramic show good contrast with reactive sputtering. Reflectivity differences between the individual components can be increased or decreased depending on the choice of the coating material.

## 2.8 Preparation and Etch Recipes of Ceramic Materials

Tables 2.1 to 2.3 show standard procedures for the preparation of dense and porous ceramics as well as for metal ceramics. These methods can be applied with flexibility of pressure and time, and they can be adjusted according to the specific material. They are basic guidelines and can be applied for the preparation of all ceramic materials listed in sections 2.8.1 to 2.8.6.

**Table 2.1 Standard preparation for dense ceramics and hard metals (semiautomatic)**

| Preparation steps | Substrate | Grain size (diamond)(a), μm | Rotating disk speed, rpm | Pressure, N | Time, min | Lubricant/Coolant |
|---|---|---|---|---|---|---|
| 1. Grinding | Diamond disk | 64 | 300 | 180 | Until flat | Water |
| 2. Grinding | Diamond disk | 20 | 300 | 150 | 5–10 | Water |
| 3. Lapping | Lapping disk | 6 | 300 | 120 | 5–10 | Alcohol based |
| 4. Polishing | Hard synthetic cloth | 3 | 150 | 150 | 5–15 | Alcohol based |
| 5. Polishing | Cotton cloth | 1 | 150 | 90 | 2–10 | Alcohol based |
| 6. Relief polishing | Short-nap cloth | 0.3 μm alumina suspension | 150 | >60 | 0.5–5 | Water |

NOTE: For cemented carbides use only step 4. (a) Diamond grits can be used as spray with the corresponding alcohol-based lubricant or as suspension recommended for the specific lapping disk and polishing cloth. Extra lubricant can be added to the diamond suspensions.

**Table 2.2 Standard preparation for porous and brittle ceramics (semiautomatic)**

| Preparation steps | Substrate | Grain size (diamond)(a), μm | Rotating disk speed, rpm | Pressure, N | Time, min | Lubricant/coolant |
|---|---|---|---|---|---|---|
| 1. Grinding | Diamond disk | 64 | 300 | 90 | Until flat | Water |
| 2. Grinding | Diamond disk | 20 | 150 | 90 | 10 | Water |
| 3. Lapping | Lapping disk | 6 | 150 | 90 | 20 | Alcohol based |
| 4. Polishing | Hard synthetic fiber cloth | 6 | 150 | 120 | 15 | Alcohol based |
| 5. Polishing | Silk/cotton cloth | 3 | 150 | 90 | 15 | Alcohol based |
| 6. Polishing | Cotton cloth | 1 | 150 | >60 | 15 | Alcohol based |

(a) Diamond grit can be used as spray with alcohol-based lubricants or as suspension recommended for the specific lapping disk and polishing cloth. Extra lubricant can be added to the diamond suspensions.

**Table 2.3 Standard preparation for metal ceramics (semiautomatic)**

| Preparation steps | Substrate | Grain size (diamond)(a), μm | Rotating disk speed, rpm | Pressure, N | Time, min | Lubricant/ coolant |
|---|---|---|---|---|---|---|
| 1. Grinding | Diamond disk | 64 | 300 | 180 | Until flat | Water |
| 2. Grinding | Diamond disk | 20 | 300 | 150 | 2 | Water |
| 3. Lapping | Lapping disk | 6 | 300 | 120 | 10 | Alcohol based |
| 4. Polishing | Nylon cloth (hard) | 3 | 300 | 240 | 5 | Alcohol based |
| 5. Polishing | Cotton cloth | 1 | 300 | 120 | 5 | Alcohol based |
| 6. Polishing | Acid-resistant synthetic cloth | $SiO_2$ suspension | 300 | >90 | 2 | Water |

(a) Diamond grit can be used as spray with alcohol-based lubricants or as suspension recommended for the specific lapping disk and polishing cloth. Extra lubricant can be added to the diamond suspensions.

## *2.8.1 Oxide Ceramics*

### Microetching

| No. | Material | Etchant | Conditions |
|---|---|---|---|
| O m1 | Pure $Al_2O_3$<br>Grain-boundary etchant, also small grains<br><br>$Al_2O_3$ with MgO<br>Grain-boundary etchant of $Al_2O_3$<br>Partial evaporation of the spinel phase<br><br>$Al_2O_3$ with additives: CaO, MgO, $SiO_2$, $Na_2O$<br>Grain-boundary and glass-phase etchant | Thermal etch in air | 20 min at 1400 °C<br>(Etch temperature is approximately 150 °C below sintering temperature.)<br>20 min at 1350 °C<br><br>15 min at 1400 °C<br>Glass phase melts, wets the surface, and solidifies again. |
| O m2 | $Al_2O_3$<br>Grain boundaries are not uniformly revealed.<br>Spinel and glass phases are dissolved. | Phosphoric acid 85% | 5 s to 3 min at 250 °C<br>Etchant can only be used for 2 to 3 samples. |
| O m3 | $Al_2O_3$<br>Grain boundaries are uniformly revealed; small grains remain intact. | Vanadium pentoxide<br>Molten | 1 min at 900 °C<br>Use platinum crucible. Preheat sample also to 900 °C.<br>Rinse in diluted aqueous hydrochloric acid 5–10 min.<br>!!! See section 4.4 in the Appendix. |
| O m4 | Spinel phase<br>Identification of the glass phase and secondary phases | 100 mL alumina 0.3 μm<br>10 mL Murakami's solution | 10–15 min etch polishing on a hard fiber synthetic cloth |

(continued)

## Microetching (continued)

| No. | Material | Etchant | Conditions |
|---|---|---|---|
| O m5 | $Al_2O_3$<br>$CeO_2$ and $Cr_2O_3$ | Potassium hydrogen sulfate<br>Molten | 15–25 s at 300 °C<br>Melt in a platinum crucible.<br>3–10 min<br>!!! See section 4.4 in the Appendix. |
| O m6 | Na beta $Al_2O_3$ | Thermal etch in air | 10 min at 1460 °C |
| O m7 | Na beta $Al_2O_3$ | Fluoroboric acid 35% | 1–2 min at 110 °C<br>!!! See section 4.4 in the Appendix. |
| O m8 | Mullite ($Al_2O_3{\cdot}SiO_2$)<br>BeO (!!! See section 4.4 in the Appendix.)<br>MgO<br>$Ca_2Zr_{1-x}O_y$ | Hydrofluoric acid 40% | Cool hydrofluoric acid to 0 °C.<br>Swab with a cotton swab.<br>!!! See section 4.4 in the Appendix. |
| O m9 | $Al_2TiO_5$<br>Grain-boundary etchant | 10 mL glycerol 87%<br>10 mL nitric acid 65%<br>10 mL hydrofluoric acid 40% | 1 min<br>!!! See section 4.4 in the Appendix. |
| O m10 | $AmO_{2-x}$<br>$AmO_{2-x}$ with $Am_2O_3$<br>!!! See section 4.4 in the Appendix. | 50 (75) mL distilled water<br>50 (25) mL nitric acid 65% | 120 s at 80 °C<br>60–80 s at 80 °C<br>Use concentration in parenthesis. |
| O m11 | $Am_2O_3$<br>!!! See section 4.4 in the Appendix. | 15 mL distilled water<br>5 mL sulfuric acid 95–97% | 20–40 s at 40–60 °C |
| O m12 | BaO in $BaTi_3O_7$<br>Grain-boundary etchant | Thermal etch in air | 1 h at 1050 °C |
| O m13 | $BaTiO_3$ | 75 (1000) mL distilled water<br>15 (15) mL hydrochloric acid 32%<br>10 (5) mL hydrofluoric acid 40% | 4 min<br>!!! See section 4.4 in the Appendix. |
| O m14 | BeO<br>!!! See section 4.4 in the Appendix. | 100 mL distilled water<br>25–50 g ammonium hydrogen fluoride | 3–5 min at 60 °C<br>Preheat sample in water.<br>!!! See section 4.4 in the Appendix. |
| O m15 | BeO<br>!!! See section 4.4 in the Appendix. | 90 mL lactic acid 90%<br>15 mL nitric acid 65%<br>5 mL hydrofluoric acid 40% | 10 min to 2 h at 65 °C<br>!!! See section 4.4 in the Appendix. |
| O m16 | CaO<br>MgO | Hydrochloric acid 32% | 3 s to 6 min |
| O m17 | CaO | Saturated aqueous sodium sulfide solution | 15–70 s<br>!!! See section 4.4 in the Appendix. |
| O m18 | $CaTiO_3$ | 75 mL distilled water<br>15 mL hydrochloric acid 32%<br>10 mL hydrofluoric acid 40% | 4 min<br>!!! See section 4.4 in the Appendix. |
| O m19 | $CeO_2$ | 50 mL distilled water<br>45 mL nitric acid 65%<br>5 mL hydrofluoric acid 40% | 15 min<br>!!! See section 4.4 in the Appendix. |
| O m20 | $CeO_2$ | (a) 10 mL distilled water<br>10 mL hydrofluoric acid 40%<br>(b) 10 mL distilled water<br>10 mL hydrochloric acid 32% | 15 min<br>First in (a) and then in (b)<br>1–2 s, moving the sample to remove etch residue<br>!!! See section 4.4 in the Appendix. |

(continued)

## Microetching (continued)

| No. | Material | | Etchant | Conditions |
|---|---|---|---|---|
| O m21 | CoO | 70 mL<br>20 mL<br>10 mL | glycerol 87%<br>hydrochloric acid 32%<br>nitric acid 65% | 30 s<br>!!! See section 4.4 in the Appendix. |
| O m22 | $Eu_2O_3$ | 10 mL<br>10 mL<br>10 mL | distilled water<br>hydrochloric acid 32%<br>saturated aqueous copper (II) sulfate solution | 5 s |
| O m23 | MgO<br>$UO_2$ (!!!)<br>!!! See section 4.4 in the Appendix. | 100 (10) mL<br>15 (10) mL | distilled water<br>nitric acid 65% | 3 s to 5 min at 25–60 °C |
| O m24 | $Nd_2O_3$<br>$U_3O_8$(!!!)<br>!!! See section 4.4 in the Appendix. | 10 (10) mL<br>1 (10) mL | distilled water<br>sulfuric acid 95–97% | 10 s<br>1–5 min at 30 °C<br>Use concentration in parenthesis. |
| O m25 | NiO | 75 mL<br>25 mL<br>1.5 mL | glacial acetic acid<br>nitric acid 65%<br>hydrofluoric acid 40% | 1 min at 80 °C<br>!!! See section 4.4 in the Appendix. |
| O m26 | $PuO_2$<br>!!! See section 4.4 in the Appendix. | 100 mL<br>8 mL | distilled water<br>sulfuric acid 95–97% | 5 s to 5 min<br>Use boiling. |
| O m27 | $UO_2$-$PuO_2$ mixtures<br>!!! See section 4.4 in the Appendix. | 80 mL<br>20 mL<br>3 drops<br>2 g | distilled water<br>nitric acid 65%<br>hydrofluoric acid 40%<br>cerium (IV) nitrate | Seconds to minutes<br>!!! See section 4.4 in the Appendix. |
| O m28 | $UO_2$, $Th_xU_yO_2$<br>$PuO_2$, as cast<br>$PuO_2$ (gamma, sintered)<br>!!! See section 4.4 in the Appendix. | 10 (1) mL<br>50 (30) mL | hydrofluoric acid 40%<br>nitric acid 65%<br>(Variable concentration) | 5–10 min at 60–80 °C<br>10 min, boiling<br>30 s to 1 min boiling<br>!!! See section 4.4 in the Appendix. |
| O m29 | PZT (PbZrTiO) | 95 mL<br>35 mL<br>2 mL | distilled water<br>hydrochloric acid 32%<br>hydrofluoric acid 40% | 10 s |
| O m30 | $SrTiO_3$<br>$MgTiO_3$ | 20 (10) mL<br>20 (9) mL<br>10 (1) mL | distilled water<br>nitric acid 65%<br>hydrofluoric acid 40% | Seconds to minutes<br>!!! See section 4.4 in the Appendix. |
| O m31 | $ThO_2$-$Y_2O_3$ mixtures<br>!!! See section 4.4 in the Appendix. | 10 mL<br>10 mL | distilled water<br>hydrochloric acid 32% | Seconds to minutes<br>Use boiling. |
| O m32 | $TiO_2$ | | Thermal etch in air | 10 min at 850 °C |
| O m33 | $TiO_2$ | 100 mL<br>5 g<br>4 mL | distilled water<br>ammonium hydrogen fluoride<br>hydrochloric acid 32% | Seconds to minutes<br>!!! See section 4.4 in the Appendix. |
| O m34 | $UO_2$<br>$UO_2$-$CeO_2$, $UO_2$-$U_4O_9$, and $UO_2$-$PuO_2$ mixtures<br>!!! See section 4.4 in the Appendix. | 1 (10) mL<br>10 (1) mL | hydrogen peroxide 30%<br>sulfuric acid 95–97% | 1–11 min<br>!!! See section 4.4 in the Appendix. |

(continued)

## Microetching (continued)

| No. | Material | Etchant | Conditions |
|---|---|---|---|
| O m35 | $U_4O_9$ !!! See section 4.4 in the Appendix. | 9 mL distilled water<br>2 mL hydrogen peroxide 30%<br>1 mL nitric acid 65% | Seconds to minutes<br>!!! See section 4.4 in the Appendix. |
| O m36 | ZnO<br>Reveals twins | 20 mL distilled water<br>1 mL glacial acetic acid | Seconds to minutes |
| O m37 | ZnO grain boundaries | 100 mL distilled water<br>10 g sodium hydroxide<br>10 g potassium ferricyanide | Seconds to minutes |
| O m38 | $ZrO_2$-TZP<br>($Y_2O_3$, RE oxide, $TiO_2$)<br>$ZrO_2$-PSZ<br>($Y_2O_3$, MgO, CaO)<br>$ZrO_2$-CSZ<br>($ZrO_2$, MgO, CaO)<br>$ZrO_2$-ZTC<br>(Possible matrices: $Al_2O_3$, mullite) | Thermal etch in air | Minutes to 1 h, 1300–1400 °C |
| O m39 | $ZrO_2$-TZP<br>$ZrO_2$-PSZ<br>$ZrO_2$-CSZ<br>$ZrO_2$-ZTC | phosphoric acid 85% | 3 s to 2 min at 250 °C<br>!!! See section 4.4 in the Appendix. |
| O m40 | $ZrO_2$ with MgO, CaO | ion beam etching | 30 min at 1–7 kV<br>60–400 mA<br>Ar/Kr atmosphere<br>Beam angle: 5–60° |
| O m41 | $ZrO_2$<br>$Zr_2O_3$ | hydrofluoric acid 40% | 2 s to 6 min<br>!!! See section 4.4 in the Appendix. |
| O m42 | $ZrO_2$ | 50 mL distilled water<br>50 mL sulfuric acid 95–97% | 1–5 min<br>Use boiling. |

## Electrolytic Etching

| No. | Material | Etchant | Conditions |
|---|---|---|---|
| O Em1 | Nb oxide<br>NbO (blue)<br>$NbO_2$ (blue green)<br>$Nb_2O_5$ (red brown) | 35 mL distilled water<br>60 mL ethanol 96%<br>30 mL saturated aqueous oxalic acid solution<br>30 mL saturated aqueous citric acid solution<br>10 mL lactic acid 90%<br>5 mL phosphoric acid 85% | 30 s to 6 min<br>17–20 V DC<br>Stainless steel cathode<br>!!! See section 4.4 in the Appendix. |
| O Em2 | NiO | 60–70 mL distilled water<br>25 mL hydrofluoric acid 40%<br>25 mL glacial acetic acid | 30–45 s<br>2–4 $mA/cm^2$<br>6–12 V DC<br>Stainless steel cathode<br>!!! See section 4.4 in the Appendix. |
| O Em3 | $UO_2$<br>!!! See section 4.4 in the Appendix | 15 mL distilled water<br>20 mL glacial acetic acid<br>10 mL sulfuric acid 95–97%<br>1 g chromium (VI) oxide | Polish 30–50 s<br>Etch 60–90 s<br>10–15 V DC<br>1 $A/cm^2$<br>Stainless steel cathode<br>!!! See section 4.4 in the Appendix. |

## 2.8.2 *Nitride Ceramics*

### Microetching

| No. | Material | Etchant | Conditions |
|---|---|---|---|
| N m1 | $AlN-La_2O_3$ | 50 mL distilled water<br>50 mL saturated aqueous sodium hydroxide solution<br>50 mL cold saturated aqueous picric acid | 1–3 min, 100 °C<br>!!! See section 4.4 in the Appendix. |
| N m2 | (Al, Ti)N,<br>$AlN-Y_2O_3$<br>TiN | 10 mL distilled water<br>10 mL glacial acetic acid<br>10 mL nitric acid 65% | Seconds to minutes<br>40–60 min at 100 °C for AlN |
| N m3 | $AlN-Al_2O_3$ | 50 mL distilled water<br>10 mL 10% aqueous sodium hydroxide solution<br>10 mL 10% aqueous potassium ferricyanide solution<br>(Modified Murakami's etchant) | 30 min at 100 °C |
| N m4 | $Si_3N_4$ | Molten sodium hydroxide (free of water) | 20 s to 3 min at 300–350°C<br>2–3 samples per melt<br>!!! See section 4.4 in the Appendix. |
| N m5 | $Si_3N_4$, UN<br>!!! See section 4.4 in the Appendix. | Thermal etching under vacuum | 15 min at 1250 °C<br>Pressure less than $1.33 \times 10^{-3}$ mbar |
| N m6 | $Si_3N_4$ | 95.4 g potassium carbonate<br>12.0 g sodium fluoride<br>Melt | 1–4 min at 900 °C<br>!!! See section 4.4 in the Appendix. |
| N m7 | $Si_3N_4$ | Hydrofluoric acid 40% | 10–15 min<br>!!! See section 4.4 in the Appendix. |
| N m8 | $Si_3N_4$ | Phosphoric acid 85% | 5–30 min at 250 °C |
| N m9 | $Si_3N_4$ | 12 g potassium hydroxide<br>12 g sodium hydroxide<br>3 g lithium hydroxide<br>Melt | 20 min at 200 °C<br>!!! See section 4.4 in the Appendix. |
| N m10 | $Si_3N_4$ | Plasma etching with $CF_4$ gas and $O_2$ gas (2:1) | 3–5 min<br>60–80 W |
| N m11 | TaN | 10 mL hydrofluoric acid 40%<br>10 mL nitric acid 65% | Seconds to minutes<br>!!! See section 4.4 in the Appendix. |
| N m12 | ZrN | 30 (60) mL glycerol 87%<br>10 (10) mL nitric acid 65%<br>10 (20) mL hydrofluoric acid 40% | Seconds to minutes<br>Add glycerol just before etching.<br>!!! See section 4.4 in the Appendix. |

### Electrolytic Etching

| No. | Material | Etchant | Conditions |
|---|---|---|---|
| N Em1 | NbN (yellow)<br>$Nb_2N$ (pink) | 35 mL distilled water<br>60 mL ethanol 96%<br>10 mL lactic acid 90%<br>5 mL phosphoric acid 85%<br>3 mL saturated aqueous oxalic acid solution<br>3 mL saturated aqueous citric acid solution | Seconds to minutes<br>17–20 V DC, stainless steel cathode<br>!!! See section 4.4 in the Appendix. |

## 2.8.3 Boride Ceramics

### Microetching

| No. | Material | | Etchant | Conditions |
|---|---|---|---|---|
| B m1 | $CrB_2$, $MoB_2$ | 10 mL<br>10 mL | hydrochloric acid 32%<br>nitric acid 65% | 1–5 min at 40 °C<br>Steam etching<br>!!! See section 4.4 in the Appendix. |
| B m2 | $HfB_2$-$NbB_2$ | 6 mL<br>2 mL<br>1 mL | hydrochloric acid 32%<br>nitric acid 65%<br>hydrofluoric acid 40% | 9 s at 30–40 °C<br>!!! See section 4.4 in the Appendix. |
| B m3 | $HfB_2$ with $TaB_2$<br>$ZrB_2$ | | Nitric acid 65% | 1–2 min at 40–50 °C<br>8–10 min |
| B m4 | $TiB_2$ | 10 mL<br>1 mL | distilled water<br>sulfuric acid 95–97% | 15 s |
| B m5 | $TiB_2$, $ZrB_2$ | 30 mL<br>10 mL<br>10 mL | lactic acid 90%<br>nitric acid 65%<br>hydrofluoric acid 40% | 5–10 s<br>Lactic acid can be replaced with 10 mL glycerol 87%<br>!!! See section 4.4 in the Appendix. |
| B m6 | $TiB_2$ | 100 mL<br>10 mL | hydrochloric acid 32%<br>nitric acid 65% | 15 s |
| B m7 | $TiB_2$ | 10 mL<br>10 mL<br>3 mL<br>3 drops | glycerol 87%<br>nitric acid 65%<br>hydrochloric acid 32%<br>hydrofluoric acid 40% | 15–30 s<br>Add glycerol just before use.<br>!!! See section 4.4 in the Appendix. |
| B m8 | $ZrB_2$ | 10 mL<br>10 mL<br>10 mL | distilled water<br>nitric acid 65%<br>hydrofluoric acid 40% | Seconds to minutes<br>!!! See section 4.4 in the Appendix. |

### Electrolytic Etching

| No. | Material | | Etchant | Conditions |
|---|---|---|---|---|
| BE M1 | $LaB_6$, TaB2 | 10 mL<br>1–2 g | distilled water<br>sodium hydroxide | Seconds to minutes<br>10–15 V DC. Stainless steel cathode |
| BE M2 | $TiB_2$ | 1000 mL<br>10 g | distilled water<br>sodium hydroxide | 30–60 s<br>30–60 V DC. Stainless steel cathode |

## 2.8.4 Carbide Ceramics

### Microetching

| No. | Material | | Etchant | Conditions |
|---|---|---|---|---|
| C m1 | Cr and Hf carbides | 10 mL<br>10 mL<br>10 mL | distilled water<br>nitric acid 65%<br>hydrofluoric acid 40% | Seconds to minutes<br>Etching time of TiC varies with its grain size. |
| | (Al,Ti)C, VC | | Use glycerol 87% instead of distilled water | !!! See section 4.4 in the Appendix. |

(continued)

## Microetching (continued)

| No. | Material | | Etchant | Conditions |
|---|---|---|---|---|
| C m2 | Cr and Mo carbides | 100 mL<br>10 g<br>10 g | distilled water<br>sodium hydroxide<br>potassium ferricyanide | 2–20 min at 50 °C |
| C m3 | SiC | | Thermal etching under vacuum | 1–3 h at 1300–1500 °C<br>Less than $1.33 \times 10^{-3}$ mbar |
| C m4 | SiC | | Sodium bicarbonate or potassium bicarbonate or sodium tetraborate<br>Melt | 10 min<br>!!! See section 4.4 in the Appendix. |
| C m5 | SSiC<br>Alpha/alpha grain boundaries<br>Alpha/beta phase boundaries | 60 mL<br>30 g<br>3 g | distilled water<br>potassium ferricyanide<br>sodium hydroxide | 5–30 min at 110 °C |
| C m6 | SSiC doped with B | 20 mL<br>10 g<br>10 g | distilled water<br>sodium hydroxide<br>potassium ferricyanide | 5–10 min |
| C m7 | SSiC doped with Al | 80 mL<br>10 g<br>10 g | distilled water<br>sodium hydroxide<br>potassium ferricyanide | 10–20 min |
| C m8 | SSiCBeta/beta grain boundaries | 90 g<br>10 g | potassium hydroxide<br>potassium nitrate<br>Melt | 5–10 min at 480 °C<br>Preheat the samples.<br>!!! See section 4.4 in the Appendix. |
| C m9 | SiC with 1% $B_4C$<br>B doped SiC | 100 mL<br>10 mL<br>10 g | distilled water<br>nitric acid 65%<br>potassium ferricyanide | 30–40 min at 60 °C<br>Seconds to minutes. Boiling |
| C m10 | SiSiC | 60 mL<br>3 g<br>30 g | distilled water<br>sodium hydroxide<br>potassium ferricyanide | 8–15 min<br>Use boiling. |
| C m11 | SiC with 5–15% oxide additions | | Plasma etching with $CF_4$ gas and $O_2$ gas (1:1) | 3–3.5 min<br>60–80 W |
| C m12 | TaC<br>(Fe,Si)C | 10 (10) mL<br>30 (10) mL | hydrofluoric acid 40%<br>nitric acid 65% | 2 min<br>Use the amounts in parenthesis for (Fe,Si)C.<br>!!! See section 4.4 in the Appendix. |
| C m13 | TaC | 10 mL<br>10 mL<br>10 mL | nitric acid 65%<br>hydrochloric acid 32%<br>sulfuric acid 95–97% | Seconds to minutes<br>Hydrofluoric acid 40% can replace the hydrochloric acid.<br>!!! See section 4.4 in the Appendix. |
| C m14 | Ta(C,N,O) | 20 mL<br>10 mL<br>10 mL | lactic acid 90%<br>nitric acid 65%<br>hydrofluoric acid 40% | Seconds to minutes<br>!!! See section 4.4 in the Appendix. |
| C m15 | (Th,U)C with a U/Th ratio of less than $3ThC_2$<br>!!! See section 4.4 in the Appendix. | 10 mL<br>10 mL | distilled water<br>nitric acid 65% | Seconds to minutes<br>1–45 min for $ThC_2$ |
| C m16 | ThC<br>!!! See section 4.4 in the Appendix. | 25 mL<br>25 mL<br>1 mL | distilled water<br>nitric acid 65%<br>hydrochloric acid 32% | Seconds to minutes |

(contined)

## Microetching (continued)

| No. | Material | Etchant | Conditions |
|---|---|---|---|
| C m17 | UC, U(C,O), U(C,N)<br>UC-ZrC mixtures<br>$UC-UC_2$, $UC_2-U_2C_3$ mixtures<br>(U,Pu)C C-rich<br>!!! See section 4.4 in the Appendix. | 10 mL distilled water<br>10 mL glacial acetic acid<br>10 mL nitric acid 65% | Seconds to minutes<br>Swabbing<br>If necessary, add some drops of aqueous oxalic acid solution or hydrofluoric acid 40%.<br>!!! See section 4.4 in the Appendix. |
| C m18 | (U,Pu)C<br>!!! See section 4.4 in the Appendix. | 8 mL ethylene glycol<br>5 mL ethanol 96%<br>5 mL phosphoric acid 85% | 5–60 s<br>Swabbing<br>!!! See section 4.4 in the Appendix. |
| C m19 | WC | 10 mL hydrochloric acid 32%<br>10 mL hydrogen peroxide 30% | Seconds to minutes<br>!!! See section 4.4 in the Appendix. |
| C m20 | ZrC | 15 mL distilled water<br>30 mL hydrofluoric acid 40%<br>15 mL nitric acid 65%<br>15 mL glacial acetic acid | 5–30 s<br>!!! See section 4.4 in the Appendix. |
| C m21 | ZrC | 100 mL distilled water<br>4 mL hydrofluoric acid 40%<br>3 g ammonium hydrogen fluoride | Seconds to minutes at 80 °C<br>!!! See section 4.4 in the Appendix. |

## Electrolytic Etching

| No. | Material | Etchant | Conditions |
|---|---|---|---|
| C Em1 | $B_4C$ | 100 mL distilled water<br>1 g potassium hydroxide | 30–60 s. 40 V DC at 3 $A/cm^2$<br>Stainless steel cathode |
| C Em2 | $B_4C$ | 100 mL distilled water<br>50 mL sulfuric acid 95–97% | 1–5 min. 4–10 V DC<br>Stainless steel cathode |
| C Em3 | $NbC/NbC_2$ | 50 mL lactic acid 90%<br>30 mL nitric acid 65%<br>8 mL hydrofluoric acid 40% | Seconds to minutes. 17–20 V DC<br>Stainless steel cathode<br>!!! See section 4.4 in the Appendix. |
| C Em4 | PuC<br>!!! See section 4.4 in the Appendix. | 7 mL distilled water<br>133 mL glacial acetic acid<br>25 g chromium (VI) oxide | 1 min. 20 V DC<br>Stainless steel cathode<br>!!! See section 4.4 in the Appendix. |
| C Em5 | PuC<br>!!! See section 4.4 in the Appendix. | 10 mL distilled water<br>10 mL nitric acid 65%<br>10 mL hydrofluoric acid 40% | 50 s. 80 $mA/cm^2$<br>Stainless steel cathode<br>PuC (dendritic)<br>!!! See section 4.4 in the Appendix. |
| C Em6 | TaC, TiC<br>SiC | 100 mL distilled water<br>2 g potassium hydroxide | 2–30 s<br>2 V DC. 30–60 $mA/cm^2$<br><br>20 s<br>6 V DC. 1 $A/cm^2$<br>Pt cathode. Keep sample agitated. |
| C Em7 | SiC<br>Grain orientation<br>Color etching | 380 mL distilled water<br>400 mL glacial acetic acid<br>30 mL sulfuric acid 95–97%<br>2 mL ethanol<br>120 g chromium (VI) oxide | 6 s<br>6 V DC. 0.4 $A/cm^2$<br>Stainless steel cathode<br>!!! See section 4.4 in the Appendix. |

(continued)

## Electrolytic Etching (continued)

| No. | Material | Etchant | Conditions |
|---|---|---|---|
| C<br>Em8 | SiC reaction bonded | 100 mL distilled water<br>10 g oxalic acid | 30 s<br>5–10 V DC. 1 A/cm$^2$<br>!!! See section 4.4 in the Appendix. |
| C<br>Em9 | SiSiC<br>Color etchant | 100 mL distilled water<br>10 g chromium (VI) oxide and a few drops of sulfuric acid 95–97% | 15 s<br>15 V DC. 1–2 A/cm$^2$<br>!!! See section 4.4 in the Appendix. |

# *2.8.5 Phosphide and Sulfide Ceramics*

## Microetching

| No. | Material | Etchant | Conditions |
|---|---|---|---|
| P<br>m1 | CdS | Hydrochloric acid 32% | 6–12 s. Boiling<br>Steam etching |
| P<br>m2 | PbS | 30 mL distilled water<br>10 mL hydrochloric acid<br>1 g 1.3-dimethyl, 2-thiourea | 1–10 min at 60 °C |
| P<br>m3 | PuP, PuS<br>!!! See section 4.4 in the Appendix. | Lactic acid 90% | Seconds to minutes |
| P<br>m4 | UP<br>!!! See section 4.4 in the Appendix. | 10 mL distilled water<br>10 mL hydrofluoric acid 40%<br>20 mL nitric acid 65%<br>20 mL sulfuric acid 95–97% | Seconds to minutes<br>!!! See section 4.4 in the Appendix. |
| P<br>m5 | US<br>!!! See section 4.4 in the Appendix. | 10 mL hydrogen peroxide 30%<br>1 mL sulfuric acid 95–97% | Seconds to minutes<br>!!! See section 4.4 in the Appendix. |

# *2.8.6 Hardmetals (Cemented Carbides)*

## Preparation

Hardmetals are liquid-phase-sintered composite materials. They consist of carbides from transition metals (W and/or Ti, Ta, Nb, and Hf) and a tough binder phase (Co, Ni, and Fe). The microstructures can be very fine grained (less than 1 μm). They are used as cutting tools; as tools for stamping, deep drawing, and wire drawing; and for abrasion-resistant parts and for machine parts. Cutting tools and abrasion-resistant parts are additionally coated with TiC, TiN, or TiCN to improve their abrasion resistance.

Flat samples, without breakouts and with a true porosity content, are obtained by using automated sample preparation equipment as described in section 2.5.2 in this chapter. However, good results can also be obtained by the following manual preparation:

*Grinding:*

Coarse grinding and fine grinding are performed on diamond grinding disks. Lapping is recommended on a lapping disk with a 15 μm diamond suspension.

*Polishing:*

On a hard synthetic cloth with a 6 and 1 μm diamond suspension and preferably without using a lubricant.

Do not use alumina for hardmetals with Co binders; the samples corrode.

## Microetching

| No. | Material | Etchant | | Conditions |
|---|---|---|---|---|
| WC m1 | WC-Ti-Ta. NbC-Co<br>Carbides are easily distinguished from the Co matrix.<br>Color etching | Thermal etching in air | | 30 min<br>400–450 °C |
| WC m2 | WC-Co, Eta phase appears multicolored; Co is white.<br>WC-TiC-TaC-Co<br>WC-NbC-Co | 100 mL<br>10 g<br>10 g | distilled water<br>potassium hydroxide<br>potassium ferricyanide<br>(Murakami's reagent) | 1 to max. 2 s to reveal the eta phase; a longer etching will attack the solid-solution phase.<br>5–20 s for mixed carbides<br>1–3 min for WC<br>Re-etch mixed carbide containing hardmetals with aqueous hydrochloric acid solution (1:1). |
| WC m3 | WC-Co, Co turns black. | | hydrochloric acid 32%, saturated with iron (III) chloride | 1–5 min |
| WC m4 | WC-Co | 100 mL<br>5 mL | hydrochloric acid 32%<br>hydrogen peroxide 30% | 60 s<br>Use only fresh! |
| WC m5 | WC-Co | 100 mL<br>20–30 g | distilled water<br>ammonium persulfate | Seconds to minutes |
| WC m6 | WC-Co<br>WC-TiC-TaC-Co<br>Grain-boundary etchant | 15 mL<br>60 (30) mL<br>15 mL<br>15 mL | distilled water<br>hydrochloric acid 32%<br>nitric acid 65%<br>glacial acetic acid | 5–30 s<br>Let stand for 1 h before use.<br>!!! See section 4.4 in the Appendix. |
| WC m7 | WC-$Mo_2C$-TiC-Ni | 10 mL<br>10 mL | methanol 99.8%<br>nitric acid 65% | 5 min<br>!!! See section 4.4 in the Appendix. |
| WC m8 | WC-TiC-NbC-Co | 75 mL<br>25 mL | hydrochloric acid 32%<br>nitric acid 65% | Seconds to 5 min<br>Use only fresh!<br>!!! See section 4.4 in the Appendix. |
| WC m9 | WC-Co or Ni matrix with $Mo_2C$, TiC, $WC_2$, TaC | 97 mL<br>3 mL | distilled water<br>hydrogen peroxide 30% | Seconds to minutes<br>Use boiling. |
| WC m10 | WC | 10 mL<br>10 mL | hydrochloric acid 32%<br>hydrogen peroxide 30% | Up to 60 s |

(continued)

## Microetching (continued)

| No. | Material | Etchant | Conditions |
|---|---|---|---|
| WC m11 | WC | 10 mL distilled water<br>10 mL nitric acid 65%<br>10 mL hydrofluoric acid 40% | Seconds to minutes<br>!!! See section 4.4 in the Appendix. |
| WC m12 | Can be used for all compositions<br>Color etchant | 100 (100) mL distilled water<br>30 (20) mL hydrochloric acid 32%<br>2 (3) mL selenic acid 96% | 15–20 min<br>Surface will change to brownish, light bluish. Final polish must be with alumina.<br>!!! See section 4.4 in the Appendix. |

## Electrolytic Etching

| No. | Material | Etchant | Conditions |
|---|---|---|---|
| WC Em1 | WC-Co<br>Good etchant for WC crystals | 100 mL distilled water<br>3 g potassium hydroxide<br>2 g sodium carbonate | 10 s<br>3 V DC. Stainless steel cathode<br>Higher sodium carbonate content will increase the dissolution of the WC phase.<br>Remove the white precipitate with diluted aqueous hydrochloric acid. |
| WC Em2 | WC-Co<br>Reveals Co | 35 mL distilled water<br>5 mL nitric acid 65% | 10 s<br>3 V DC. Stainless steel cathode |
| WC Em3 | WC-Co<br>(Suited for SEM examination) | 75 mL distilled water<br>25 mL ethanol 96%<br>3 g sodium carbonate | Seconds to minutes<br>2.5 V DC. 1 A/cm$^2$<br>Stainless steel cathode |

## Physical Etching

The optical properties of phases in hardmetals vary greatly. A good distinction of the carbides in hardmetals with solid-solution carbides can be obtained by interference layer deposition (vapor deposition of ZnSe, CdS, ZnFe, and InP, or reactive sputtering with $Fe/O_2$). This method also gives very good results when etching coated materials.

# 3 Plastography

Similar to metallography and ceramography, plastography is the examination of microstructures of plastic materials or polymers. Polymers are macromolecular organic materials consisting of base molecules (monomers) that are aligned in long molecular strings (macromolecules), forming networks or chains. They may be produced synthetically from raw materials such as crude oil, natural gas, or coal, but also by transformation of the original macromolecular natural products such as cellulose, protein, or natural rubber.

The variety of the different polymers is a consequence of the chemical reaction (polymerization, polycondensation, polyaddition), and the resulting arrangement of the macromolecules (linear cross linked: thermoplastics; spatially and coarsely linked: elastomers; spatially and densely cross linked: duromers). On the other hand, many other variations can be obtained by the size of the macromolecules (molecular weight or degree of polymerization) and their molecular arrangements (partial crystalline structures, spherolites, more or less strong orientations in one or two directions), all having an effect on a wide range of properties.

Typical properties of polymers are their low density, easy formability, low heat conductivity, limited strength at elevated temperatures, large heat expansion, and good chemical resistance for inorganic substances. Their mechanical properties such as strength, toughness, elasticity, hardness, and so forth, vary with their composition. The wide range of composition and properties of polymers opened large possibilities of their applications. Table 3.1 shows the different types and groups of plastic materials and their abbreviations.

Microstructures of polymer samples can be examined with reflective light as can metallographic or ceramographic samples. However, microscopic examination of thin sections using transmitted light is more common. The choice depends on the material and the purpose of the examination. A list of various choices for the preparation and examination methods is shown in Table 3.2.

## Table 3.1 Typical examples of plastic materials and their abbreviations

| Type of plastic material | Name of plastic material | Abbreviation |
|---|---|---|
| Amorphous thermoplastics | Acrylonitrile-butadiene-styrene | ABS |
| | Polyamide | PA |
| | Polycarbonate | PC |
| | Polymethyl methacrylate | PMMA |
| | Polyphenylene oxide | PPO |
| | Polystyrene | PS |
| | Polyvinyl chloride | PVC |
| | Styrene acrylonitrile copolymer | SAN |
| Partial crystalline thermoplastics | Polyamide | PA |
| | High-density polyethylene | HDPE |
| | Polybutadiene | PB |
| | Polyethylene | PE |
| | Polyethylene terephthalate | PETP |
| | Polyoxymethylene | POM |
| | Polypropylene | PP |
| | Polytetrafluoroethylene | PTFE |
| Duroplastic materials | Epoxy resin | EP |
| | Phenolic resin | PF |
| | Polyurethane | PUR |
| Elastomers | Alkylene sulfide elastomer | ASM |
| | Butadiene elastomer | BR |
| | Isoprene rubber | IR = NR |
| | Isoprene-isobutylene rubber | IIR |
| | Natural rubber | NR |
| Liquid-spun fiber (not meltable) | Polyacrylonitrile | PAN |
| High-temperature-resistant plastic material | Polyimide | PI |

## Table 3.2 Preparation and evaluation criteria of polymers for light microscopic observations

| Preparation/examination method | Purpose of examination | Material |
|---|---|---|
| Surface preparation<br>(a) Grinding and polishing<br>(b) Ultramilling<br>Etching<br>Reflective light observation (brightfield, darkfield, differential interference contrast) | Large sample surface, fabrication defects, flow lines, binding defects, spherolite size, evaluation of filler material portion, and phase bonding | Elastomers, duromers, two-phase systems, e.g., duroplastics with fillers (glass fibers, metals, wood, etc.), layered composites<br>Polyamide, polypropylene, polyoxymethylene are easier to prepare by ultramilling. |
| Thin sections:<br>(a) Grinding and polishing on both sides<br>Transmitted light (brightfield and polarized transmitted light) | Distribution of fibers, spherolite structures | Duromers, partial crystalline thermoplastics, hard, brittle, or duroplastics containing nonmetallic filler material |
| Microtome sections:<br>Transmitted light (brightfield and polarized transmitted light)<br>Freezer sectioning | Superstructures, flow lines, amorphous and crystalline regions, size and distribution of spherolites | Amorphous thermoplastics, partial crystalline thermoplastics, foam materials<br>Polyethylene: soft and hard |

## 3.1 Plastographic Sample Preparation

The same steps of metallographic sample preparation apply in principle in plastography: sampling, mounting, identification, grinding and polishing, cleaning, and etching. Naturally, certain exceptions do apply. In either case, certain steps may be eliminated.

**Sampling.** Only mechanical methods are used. Large sections can be cut with a bandsaw or hacksaw. A low-speed diamond saw with a hard diamond blade is recommended for precision cutting. Fiber-reinforced polymers are best cut with a soft diamond blade, and running water is used as a coolant. It is important to cut at a sufficient distance away from the area to be examined. When cutting too close to the area to be analyzed, microcracks may be generated, which can lead to misinterpretation.

**Mounting.** Of the mounting methods listed in Fig. 1.2 (in Chapter 1), only room-temperature curing or cold-setting materials (see Table 1.2 in Chapter 1) can be used since the tolerance for elevated temperature of some polymers is relatively low. The selection of an appropriate mounting material (which is also a polymer) must take into consideration its hardening temperature, which has to be compatible with the softening temperature of the polymer sample. Table 3.3 lists the softening temperatures of some polymers. The heat generated during the hardening process can be reduced by mounting in several steps and by including intermediate cooling. Successive mounting steps are also necessary when the sample material has a lower density than the mounting material. Glass fiber reinforced polymers also require a harder filler material, which is added to the mounting material for better hardness adaptation.

**Identification.** The same identification methods apply for plastographic samples as listed in section 1.4 (in Chapter 1).

**Table 3.3 Softening and maximum high-temperature use of some plastic materials**

| Polymers | Softening temperature, °C | Maximum high-temperature use, °C |
|---|---|---|
| Polyvinyl chloride | 70–90 | 60–75 |
| Polystyrene | 75–105 | 60–80 |
| Polypropylene | 80–90 | 90–110 |
| Acrylonitrile-butadiene-styrene | 85–105 | 85–105 |
| Polymethyl methacrylate | 85–125 | 65–95 |
| Styrene acrylonitrile copolymer | 95–105 | 85–95 |
| Polyurethane | 100–180 | 80–110 |
| Polyamide | 130–220 | 70–110 |
| Polycarbonate | 145–155 | 110–130 |
| Polyoxymethylene | 160–170 | 85–110 |

**Grinding and Polishing.** Mechanical grinding and polishing is normally used. Since polymers are elastic, they do not produce a clean chip. The grinding grooves of a polymer surface are therefore more or less roughened instead of smoothed. Soft materials tend to smear, especially after the fine grinding and polishing steps. It is very important to work with continuously sharp (fresh) cutting media to ensure good removal of the abraded material.

*Grinding* is typically performed with SiC papers from 320 up to 1200 grit using running water. *Polishing* is carried out on hard polishing cloths with diamond and with lubricants that are compatible with the plastic material. *Duroplastic materials* can usually be quite easily ground and polished; *elastomers* on the other hand require sharp cutting grains (diamond), hard polishing cloths, and long polishing times because of their high abrasion resistance and elasticity.

Table 3.4 shows the polishing conditions for both groups of plastic materials. Glass fiber reinforced polymers are also difficult to prepare; the fibers tend to pull out during grinding and polishing and may form a relief during polishing because of the hardness difference between the matrix and the fibers. These samples are best prepared with automatic equipment at a controlled sample speed during polishing, at low pressure and using hard cloths, and by continuously adding diamond particles for grinding and polishing; this is followed by a final polish on a soft low-nap cloth using an alumina, $SiO_2$, or $Cr_2O_3$ suspension.

**Table 3.4 Semiautomatic sample preparation of duroplastics and elastomers**

| Polishing steps for polymers | Substrate | Diamond grit size, μm | Speed, rpm | Load, g | Time, min |
|---|---|---|---|---|---|
| **Duroplastic(a)** | | | | | |
| 1 | Hard synthetic cloth | 6 | 150 | 500 | 5 |
| 2 | Woven cotton cloth | 3 | 150 | 500 | 10 |
| 3 | Short-nap velvet cloth | 1 | 150 | 50 | 5 |
| 4 | Short-nap velvet cloth | 0.25 | 150 | 50 | 5 |
| 5 | Short-nap velvet cloth | 0.05 $Al_2O_3$ | 150 | 50 | 5 |
| **Elastomer(b)** | | | | | |
| 1 | Hard synthetic cloth | 6 | 150 | 600 | 10 |
| 2 | Hard woven silk cloth (hard synthetic cloth) | 3 | 150 | 600 | 20 |
| 3 | Short-nap velvet cloth (hard synthetic cloth) | 1 | 150 | 600 | 20 |
| 4 | Short-nap velvet cloth | 0.05 $Al_2O_3$ | 300 | Manual | 5 |

(a) Steps 4 and 5 can be alternated. Use oil-based lubricants. (b) Cloths in parentheses are recommended for polymers reinforced with filler materials. Use water as lubricant; alcohol-containing lubricants attack elastomers.

*Ultramilling* (see section 1.5.2 in Chapter 1) provides quick sample preparation for microscopic and macroscopic observation. A milling tool with two diamond cutters removes thin layers of the sample material. Perpendicular to the tool is an oscillating holder with the sample. The cutting depth is varied by adjusting the sample holder in the direction of the milling head. The quality of the prepared surface depends largely on the cutting speed, the depth of cutting, and the knife and milling head geometry, as well as the adjustment of knife and milling head and the cooling of the sample. The individual parameters vary for different polymer materials; it is important to choose a coolant that does not dissolve the materials or produce stress cracking. Ultramilling is very successful for glass fiber reinforced plastic materials.

**Cleaning,** especially between individual steps, is very important in plastographic sample preparation. It must be noted, though, that many liquids used for rinsing metallographic and ceramographic samples will attack polymers and cannot be used in plastography. This applies especially for organic liquids. Table 3.5 lists liquids that may cause softening (partial dissolution) or swelling in plastic materials. Though they are suitable as etchants in plastographic sample preparation, they cannot be used for cleaning. Alcohol and water remain the most suitable cleaning liquids.

## 3.2 Plastographic Etching

As-polished and unetched polymer samples can be examined with reflective light or the scanning electron microscope for cracks and bonding defects between matrix and filler material. The low reflectivity of the samples can be considerably improved with a thin-film deposition of carbon or gold. To reveal the numerous other microstructural features such as phase boundaries, textures, or distribution of spherolites, the low contrast must be improved by etching. The available possibilities are mainly restricted to optical etching and darkfield, differential interference contrast, as well as chemical etching.

**Optical Etching.** Optical contrasting with reflective light uses *darkfield illu mination* to reveal small fissures and *differential interference contrast* for phase examination in relief. A deposition of a thin metal layer, for example Au (10 nm), provides better *reflectivity* and should be used when photomicrographs are required.

**Chemical Etching.** Crystalline regions in polymers can be revealed by etching with acids or by partial dissolution of the surface with solvents. Crystalline regions are attacked at a lesser rate than amorphous regions; this provides sufficient contrast of the individual microstructural features. To avoid a blurred microscopic image, care must be taken not to overetch the amorphous regions. Organic solvents are easier to handle than etching with acids and are therefore

**Table 3.5 Examples of liquids that cause partial solution or swelling in polymers (unsuitable for cleaning)**

| Plastic group | Solution causing partial solution or swelling |
|---|---|
| **Partial crystalline thermoplastics** | |
| Polyethylene | Benzine, benzene, chloroform, tetra-chloromethane, toluene, trichloroethylene |
| Polypropylene | Benzine, benzene, chloroform, toluene, trichloroethylene |
| Polyoxymethylene | Chloroform |
| Polyethylene terephthalate | Acetone, chloroform |
| **Amorphous thermoplastics** | |
| Polystyrene | Acetone, benzine, benzene, chloroform, tetra-chloromethane, toluene, trichloroethylene |
| Polyvinyl chloride | Acetone, benzene, chloroform, toluene, trichloroethylene |
| Acrylonitrile-butadiene-styrene | Acetone, benzene, chloroform, tetra-chloromethane, toluene, trichloroethylene |
| Polymethyl methacrylate | Acetone, benzene, chloroform, ethanol, tetra-chloromethane, toluene, trichloroethylene |
| Polycarbonate | Acetone, benzene, chloroform, tetra-chloromethane, toluene, ethanol, trichloroethylene |
| Epoxy resin | Acetone, chloroform, trichloroethylene |
| Polyphenylene ether | Acetone, benzine, chloroform, tetra-chloromethane |
| **Elastomers** | |
| Polyurethane | Acetone, chloroform, tetra-chloromethane, trichloroethylene |
| Isoprene-isobutylene rubber | Benzine, benzene, chloroform, tetra-chloromethane, toluene, trichloroethylene |
| Natural rubber | Benzine, benzene, chloroform, tetra-chloromethane, toluene, trichloroethylene |
| Alkylene sulfide elastomer | Acetone, tetra-chloromethane, trichloroethylene |
| **Duroplastics** | |
| | None |

preferred. Etching time and temperature can affect the microstructure of some polymers, for example, polystyrol. The corrosive attack of the etchant can reduce inherent stress within the material by stress corrosion. In fact, this effect can be used to examine stress distribution.

## 3.3 Etchants for Polymers

| No. | Material | Etchant | | Conditions |
|---|---|---|---|---|
| P m1 | Polyoxymethylene | 100 mL<br>30 mL | distilled water<br>hydrochloric acid 23% | 20 s |
| P m2 | Polyoxymethylene<br>Reveals spherolite cores and growth direction.<br>Polypropylene | | Nitric acid 65% | 10 min |

(continued)

## Etchants for Polymers (continued)

| No. | Material | Etchant | Conditions |
|---|---|---|---|
| P m3 | Polypropylene | 100 mL nitric acid 65%<br>100 mL chromosulfuric acid (Variable concentration) | 2 min at 70 °C<br>Time varies.<br>!!! See section 4.4 in the Appendix. |
| P m4 | Polypropylene (Attack is slower for amorphous regions.) | 100 mL distilled water<br>60 g chromium (VI) oxide | Several h at 70 °C<br>!!! See section 4.4 in the Appendix. |
| P m5 | Spherolites in polyethylene Polypropylene | Xylol 99.8% | 3 days at 70 °C<br>!!! See section 4.4 in the Appendix. |
| P m6 | Polyethylene (Reveals lamellar structure) | Nitric acid 65% | Seconds to minutes |
| P m7 | Polycarbonate of styrene acrylonitrile copolymer | Triethylamine 99% | Seconds to minutes<br>!!! See section 4.4 in the Appendix. |
| P m8 | Mixtures of acrylonitrile-butadiene-styrene | 100 mL distilled water<br>Add chromium (VI) oxide to saturate the solution. | Produce fracture surface in liquid $N_2$, then 5 min at 40 °C<br>!!! See section 4.4 in the Appendix. |
| P m9 | Acrylonitrile-butadiene-styrene | 10 mL distilled water<br>40 mL sulfuric acid 95-97%<br>10 mL phosphoric acid 85%<br>2 g potassium dichromate | 3 min at 70 °C<br>!!! See section 4.4 in the Appendix. |
| P m10 | Glass fiber reinforced epoxy resins | 99.5 mL distilled water<br>0.5 mL hydrofluoric acid. 40% | Few s, attacks the interface of matrix/glass fiber<br>!!! See section 4.4 in the Appendix. |
| P m11 | Glass fiber reinforced epoxy resins<br>Color etchant | 90 mL distilled water<br>10 g chromium (VI) oxide | Etch in an ultrasonic bath. Up to 10 min: fiber color is blue. Up to 30 min fiber color varies.<br>!!! See section 4.4 in the Appendix. |
| P m12 | Glass fiber reinforced polyester resins | 60 mL distilled water<br>40 mL hydrofluoric acid 40% | Seconds to minutes<br>Fibers are selectively dissolved.<br>!!! See section 4.4 in the Appendix. |

## 3.4 Plastographic Thin Sectioning

Only hard polymers and those containing nonmetallic filler materials are suited for thin sections and transmitted light microscopy. The specimens are mounted in epoxy resin under a light vacuum, and specimens with a rough surface are prepared as usual by grinding on SiC papers and then polishing on a hard synthetic cloth up to 1 μm diamond (see section 3.1 in this chapter). The flat surface is glued onto an objective glass slide with a suitable glue, for example, Canada-Balsam, epoxy resin, or a cyanacrylate glue (instant glue). The sample must be dry and free of grease. The thickness of the glue layer is 3 to 5 μm. Larger size samples with a flat and smooth surface can be glued onto the slide directly without mounting and grinding. The sample is then thinned, if possible, with a commercial thin-sectioning machine. The sample slide is fitted into a vacuum holder and is ground on a diamond disk until the required thickness is obtained (approximately 100 μm). After thinning the sample by grinding, the slide is mounted into a special thin section holder to carry out the final thinning by polishing either manually or automatically down to a final thickness of 20 to 10 μm. Suitable polishing cloths are hard silk cloths, and suitable polishing media are diamond grit sizes from 6 to 1 μm; the lubricant must be compatible with the polymer material.

## 3.5 Plastographic Microtome Sectioning

*Microtome sectioning* is more often used than the thin sectioning method by grinding and polishing. Thin slices of a uniform thickness are removed by a microtome. The thickness may be adjusted to less than 10 μm. With ultramilling (see section 1.5.2 in Chapter 1), slices thinner than 1 μm can be cut.

Presampling prior to thin sectioning is done with a saw or a jeweller's saw. Very sensitive samples can be sectioned with a diamond wire saw. The sectioned samples are then mounted into small vises on the microtome. The cutting width should be small. Intricate samples should be fixed with glue. Suitable glues are component glues or a methylacrylate glue. Small samples or fractures can be fixed by mounting in epoxy resin. If the samples are extremely soft, for example, polyethylene, and the cutting action causes deformation, the sample must be hardened by a suitable coolant. This technique, the *freezing technique,* is commonly used in biology and medicine. The samples are placed on a freezer table that is cooled with acetone/dry ice or liquid nitrogen and then cooled and sectioned. Thin sections ranging from 1 to 10 μm can be obtained this way.

In this thin-sectioning technique, the microtome knife is very important. Knives of different materials, such as hardened steel, hardmetal glass, or diamond are available. These knives are also made with different cutting edges. Specialized training and experience are necessary to choose the right combination of knife, cutting edge, cutting angle, and cutting thickness for the sample preparation. Regarding the adjustment of the cutting angle, the softer the material to be cut, the smaller the cutting angle should be. Large objects and hard materials require a large cutting angle for stability. Table 3.6 shows guidelines for microtome-cutting angles used for plastic materials.

**Table 3.6 Guidelines for microtome knife cutting angles for polymers**

| Polymers | Microtome knife cutting angle |
|---|---|
| Acrylonitrile-butadiene-styrene, styrene acrylonitrile copolymer, polyvinyl chloride | 45° |
| Polyamide, polypropylene, polyoxymethylene | 30° |
| Polyethylene | 15° |

The finished sections are usually difficult to remove from the knife. They tend to curl. They must be flattened with a pair of tweezers, a wooden spatula, some paraffin oil as a wetting agent, or some adhesive tape onto the objective slide holder. The sections can be used with a dry medium such as Canada balsam or with a liquid medium such as paraffin oil, which is placed between the objective slide and the cover glass. The sections are then ready for microscopic analysis and can also be easily stored.

Finally, another technique, the *cast liquid film technique,* should be mentioned; this technique is used for thermoplastic materials and liquid crystalline plastics. It is used primarily to characterize polymerization processes. A plastic material solution is placed onto a glass plate, the solvent is evaporated, and the remaining plastic material is examined with transmitted light.

## 3.6 Light Microscopic Studies of Thin Sections or Microtomed Sections

Microscopic examination of thin sections is carried out in transmitted light and typically in *brightfield illumination, polarized light,* and with *phase contrast.*

In brightfield illumination, the light amplitude (brightness of the light) is reduced during the transmission of white light through the absorbing object, and an image with darker regions is generated (amplitude object). If the sample contains microstructural or stress inhomogeneities, the lightways are diffused differently and will interfere with corresponding differences revealing microstructural or stress contrasts.

Improved contrasts of these samples can be obtained using *phase contrast* (see section 1.7.1 in Chapter 1), where invisible microstructural details can be accentuated by a light-dark contrast.

Optically anisotropic materials (noncubic crystal structures) display an optical double diffraction using *polarization* with cross-polarized filters (see section 1.7.1 in Chapter 1), resulting in a periodic distinction and brightening. Polymers having a crystalline or partial crystalline (spherolite) microstructure also display specific color contrasts in polarization. Stressed isotropic polymers are optically anisotropic, and their refractive index depends on the direction. It is therefore possible to detect stress conditions even in amorphous plastic materials by polarization, and conclusions can be drawn related to the production and the expected properties of the plastic materials.

# 4 Appendix

## 4.1 Designation of Etchants

In the literature, etchants are occasionally referred to by their name, but without details of their composition, for example, Murakami or Oberhoffer. Even though these designations and their composition have already been described in sections 1.9 (in Chapter 1 "Metallography") and 2.8 (in Chapter 2 "Ceramography"), they may not be quickly referenced this way. They have therefore been listed again in alphabetical order.

| Name | Composition | Application |
|---|---|---|
| **Adler**<br>Macroetchant<br>Microetchant | 25 mL distilled water<br>50 mL hydrochloric acid 32%<br>15 g iron (III) chloride<br>3 g ammonium tetrachlorocuprate (II)<br>!!! See section 4.4 in this Appendix. | High-alloy, corrosion-resistant steels<br>Co and Ni alloys |
| **Aqua regia** | 60 mL hydrochloric acid 32%<br>20 mL nitric acid 65%<br>!!! See section 4.4 in this Appendix. | Pure Au and noble metal alloys<br>Use fresh! |
| **Barker**<br>Anodic etchant | 200 mL distilled water<br>5 g fluoboric acid 35%<br>!!! See section 4.4 in this Appendix. | Pure Al and Al alloys<br>Al, Pb, or stainless steel cathode<br>20–40 V DC |
| **Baumann print**<br>Sulfur print<br>Macroetchant | 100 mL distilled water<br>5 mL sulfuric acid 95–97% | Unalloyed steels. Sulfide inclusions<br>Grind sample to grit 320 SiC paper. Soak photographic paper with the solution and press firmly onto the sample surface. After approximately 3 min rinse, fix (photo fix), wash, and dry. S-rich regions will turn brown. Better contrast is obtained by etching the sample with Heyn's reagent prior to the Baumann print. |
| **Beraha I**<br>Color etchant | 100 mL Beraha I stock solution<br>1 g potassium disulfite<br><br>**Beraha I stock solution:**<br>1000 mL distilled water<br>200 mL hydrochloric acid 32%<br>24 g ammonium hydrogen fluoride<br>!!! See section 4.4 in this Appendix. | Unalloyed and low-alloy steels. Mn hard steels<br><br><br><br>Store the stock solution in a plastic bottle. |

(continued)

# Designation of Etchants (continued)

| Name | Composition | Application |
|---|---|---|
| **Beraha II**<br>Color etchant | 100 mL Beraha II stock solution<br>1g potassium disulfite<br><br>**Beraha II stock solution:**<br>800 mL distilled water<br>400 mL hydrochloric acid 32%<br>48 g ammonium hydrogen fluoride<br>!!! See section 4.4 in this Appendix. | High-alloy steels<br><br><br><br>Store the stock solution in a plastic bottle. |
| **Beraha III**<br>Color etchant | 100 mL Beraha III stock solution<br>1 g potassium disulfite<br><br>**Beraha III stock solution:**<br>600 mL distilled water<br>400 mL hydrochloric acid 32%<br>50 g ammonium hydrogen fluoride<br>!!! See section 4.4 in this Appendix. | Ni-base and Co-base alloys<br><br><br><br>Store stock solution in a plastic bottle. |
| **Beraha III 1**<br>Color etchant | 50 mL distilled water<br>7 g ammonium hydrogen fluoride<br>50 mL hydrochloric acid 32%<br>0.5 g potassium disulfite<br>!!! See section 4.4 in this Appendix. | Ni-base alloys<br>Hastelloy X<br>Hastelloy |
| **Bitter-technique** | Solution A:<br>300 mL distilled water<br>2 g iron (II) chloride<br>5.4 g iron (III) chloride<br>Heat solution A to 30–40 °C.<br><br>Solution B:<br>50 mL distilled water<br>5 g sodium hydroxide<br><br>In order to make a colloidal $Fe_3O_4$ suspension, solution A is added slowly but by stirring vigorously to solution B. $Fe_3O_4$ is filtered and washed several times. Mix 1 $cm^3$ $Fe_3O_4$ with 30 mL of a 0.3% soap solution. Add 1 drop of this suspension to the sample surface and cover with a cover glass for uniform distribution. Examine under the microscope with brightfield or darkfield illumination. The $Fe_3O_4$ particles will decorate the domain walls because of their nonuniform magnetic fields. | Magnetic materials<br>The samples should be electrolytically polished. |
| **Carapella**<br>Macroetchant | 20–100 mL distilled water or ethanol 96%<br>2–25 mL hydrochloric acid 32%<br>5–8 g iron (III) chloride | Ni-Fe, Ni-Cu, and Ni-Ag alloys<br>Ni-base superalloys<br>Monel metal |
| **Chromosulfuric acid** | 50 g sodium dichromate or potassium dichromate or 20 g chromium (VI) oxide<br>1000 mL sulfuric acid 95–97%<br>!!! See section 4.4 in this Appendix. | Polypropylene |

(continued)

# Designation of Etchants (continued)

| Name | Composition | Application |
|---|---|---|
| **Crowell**<br>Microetchant | 100 mL distilled water<br>3 mL sulfuric acid 95–97%<br>2 g potassium dichromate<br>1 g sodium fluoride<br>!!! See section 4.4 in this Appendix. | Ag-Sn-Zn-Cu alloys<br>Cu-Sn-Ag dental alloys |
| **Czochralski**<br>Microetchant | 30 mL distilled water<br>15 mL hydrochloric acid<br>50 mL aqueous sodium thiosulfate 16%<br>3 mL aqueous chromium (VI) oxide solution 10%<br>!!! See section 4.4 in this Appendix. | Sb alloys |
| **Dix-Keller**<br>Microetchant | 190 (190) mL distilled water<br>5 (10) mL nitric acid 65%<br>3 (6) mL hydrochloric acid 32%<br>2 (4) mL hydrofluoric acid 40%<br>!!! See section 4.4 in this Appendix. | Pure Al and Al alloys |
| **Flick**<br>Macroetchant | 90 mL distilled water<br>15 mL hydrochloric acid 32%<br>10 mL hydrofluoric acid 40%<br>!!! See section 4.4 in this Appendix. | Pure Al and Al alloys |
| **Fry**<br>Macroetchant | 100 mL distilled water<br>120–180 mL hydrochloric acid 32%<br>45–90 g copper (II) chloride | Flow lines in low-carbon $N_2$ steels. Thomas steels |
| Microetchant | 30 mL distilled water<br>40 mL hydrochloric acid 32%<br>25 mL ethanol 96%<br>5 g copper (II) chloride | Anneal sample at 150–200 °C prior to etching. After etching immerse into 32% hydrochloric acid, rinse with water, and neutralize with a 25% ammonium hydroxide solution. |
| **Groesbeck**<br>Color etchant | 100 mL distilled water<br>4 g sodium hydroxide<br>4 g potassium permanganate | High-alloy steels<br>Cast materials |
| **Hasson**<br>Color etchant | 75 mL ethanol 96%<br>40–60 mL aqueous iron (III) chloride solution (1300 g/L)<br>25 mL hydrochloric acid 32% | Pure Mo |
| **Heyn**<br>Macroetchant | 100 mL distilled water<br>9 g ammonium tetrachlorocuprate II | Low-carbon steels |
| **Kalling 1**<br>Microetchant | 33 mL distilled water<br>33 mL ethanol 96%<br>33 mL hydrochloric acid 32%<br>1.5 g copper (II) chloride | Martensitic stainless steels |
| **Kalling 2**<br>Microetchant | 100 mL ethanol 96%<br>100 mL hydrochloric acid 32%<br>5 g copper (II) chloride | Stainless steels<br>Ni-Cu alloys<br>Ni-base superalloys |

(continued)

# Designation of Etchants (continued)

| Name | Composition | Application |
|---|---|---|
| **Keller**<br>Microetchant<br>Macroetchant | 190 (20) mL distilled water<br>3 (20) mL hydrochloric acid 32%<br>5 (20) mL nitric acid 65%<br>1 (5) mL hydrofluoric acid 40%<br>!!! See section 4.4 in this Appendix. | Pure Al and Al alloys<br>Ti alloys<br>RE-Al alloys |
| **Klemm** | Stock solution:<br>300 mL distilled water<br>1000 g sodium thiosulfate<br><br>Heat solution to 30–40 °C. Let stand for one day; a part of the salt will crystallize at the bottom of the beaker. Use this stock solution and replenish the remaining solution with distilled water until all the crystals have been dissolved again.<br>This stock solution can be stored indefinitely. The etchant itself can be used for 1–2 days. | |
| **Klemm I**<br>Color etchant | 100 mL stock solution<br>2 g potassium disulfite | Unalloyed and low-alloy steels. Cast iron. Mn steels. Pure Zn and low-alloy Zn |
| **Klemm II**<br>Color etchant | 100 mL stock solution<br>5 g potassium disulfite | Cu and Cu alloys and soft solders. Solders and welds with brass |
| **Klemm III**<br>Color etchant | 100 mL distilled water<br>11 mL stock solution<br>40 g potassium disulfite | Pure Cu and Cu alloys |
| **Kroll**<br>Microetchant | 100 mL distilled water<br>2–6 mL nitric acid 65%<br>1–3 mL hydrofluoric acid 40%<br>!!! See section 4.4 in this Appendix. | Al-Cu alloys. Ti alloys, mainly Ti-Al-V(-Sn) alloys |
| **Lichtenegger and Bloech**<br>Color etchant | 100 mL distilled water<br>20 g ammonium hydrogen fluoride<br>0.5 g potassium disulfite<br>!!! See section 4.4 in this Appendix. | Austenitic CrNi steels |
| **Marble**<br>Macroetchant<br>Microetchant | 100 mL distilled water<br>100 mL hydrochloric acid 32%<br>20 g copper (II) sulfate | Austenitic and high-temperature-resistant steels<br>Co superalloys |
| **Murakami**<br>Microetchant | 100 mL distilled water<br>5–10 g potassium or sodium hydroxide<br>5–10 g potassium ferricyanide | Pure W and W alloys and W-Ni alloys. Cast iron. Alloyed Cr steels. Cr, Mo, and Mo-Cr alloys. Mo-Fe and Mo-Re alloys. W and W-base alloys. Re and Re-base alloys. InP, SbTe, and BiTe<br>U alloys (!!! See section 4.4 in this Appendix.)<br>SiC<br>Use fresh! |
| **Nital**<br>Microetchant | 100 mL ethanol 96%<br>1–10 mL nitric acid 65% | Pure iron, carbon steels, low-alloy steels, and gray cast iron. Pure Mg and most Mg alloys. Pure Sn. Sn-rich alloys with Cd, Cu, Fe, and Sb. Pure Ti. Pure Cd and Cd alloys. Pure Gd. RE-Co alloys |

(continued)

# Designation of Etchants (continued)

| Name | Composition | | Application |
|---|---|---|---|
| **Oberhoffer**<br>Macroetchant | 500 mL<br>500 mL<br>50 mL<br>30 g<br>1 g<br>0.5 g | distilled water<br>ethanol 96%<br>hydrochloric acid 32%<br>iron (III) chloride<br>copper (II) chloride<br>tin (II) chloride | Unalloyed and low-alloy steels |
| **Palmerton**<br>Macroetchant | 100 mL<br>20 g<br>1.5 g | distilled water<br>chromium (VI) oxide<br>sodium sulfate anhydrous<br>!!! See section 4.4 in this Appendix. | Cu-containing Zn alloys |
| Microetchant | 200 mL<br>20 g<br>1.5 g | distilled water<br>chromium (VI) oxide<br>sodium sulfate anhydrous<br>!!! See section 4.4 in this Appendix. | Pure Zn and Zn alloys<br>Pb-containing Zn rolling alloys |
| **Picral**<br>Microetchant | 100 mL<br>4 g | ethanol 96%<br>picric acid<br>!!! See section 4.4 in this Appendix. | Pure iron, carbon steels, low-alloy steels, and cast iron<br>Sn-coated steels and cast irons |
| **Schramm**<br>Macroetchant<br>Microetchant | 900 mL<br>60 mL<br><br>122 g<br>75 g<br>6.5 g | distilled water<br>saturated aqueous copper (III) nitrate solution<br>potassium hydroxide<br>potassium cyanide<br>citric acid<br>!!! See section 4.4 in this Appendix. | Zn-Cu, Zn-Fe, Zn-Mg, Zn-Ni, and Zn-Pb alloys<br>The mixing sequence must be followed, or cyanide acid may be generated; this may also occur during incorrect disposal practice! |
| **Tucker**<br>Macroetchant | 25 mL<br>45 mL<br>15 mL<br>15 mL | distilled water<br>hydrochloric acid 32%<br>nitric acid 65%<br>hydrofluoric acid 40%<br>!!! See section 4.4 in this Appendix. | Pure Al, Mn-Al, Si-Al, Mg-Al, Mg-Si-Al alloys<br>Use fresh! |
| **Villela**<br>Microetchant | 45 mL<br>30 mL<br>15 mL | glycerol 87%<br>hydrochloric acid 32%<br>nitric acid 65%<br>(Variable concentration)<br>!!! See section 4.4 in this Appendix. | Ferritic Cr steels. Austenitic CrNi steels. Austenitic cast alloys. Pd and its alloys<br>Use fresh! |
| **Villela and Beregekoff**<br>Macroetchant | 80 mL<br>10 mL<br>10 mL | glycerol 87%<br>glacial acetic acid<br>nitric acid 65%<br>!!! See section 4.4 in this Appendix. | Pb-Sn alloys |
| **Wallner**<br>Microetchant | 16 g<br><br>9 g<br><br>60 mL | iron chloride dissolved in 50 mL distilled water<br>ammonium persulfate dissolved in 50 mL distilled water<br>hydrochloric acid 32% | Corrosion-resistant CrNi steels |
| **Weck**<br>Color etchant | 100 mL<br>50 mL<br>2 g | distilled water<br>ethanol 96%<br>ammonium hydrogen fluoride<br>!!! See section 4.4 in this Appendix. | Pure Ti and Ti alloys |

## 4.2 List of Chemicals

The chemicals listed here are those used for the etchants in sections 1.9 (in Chapter 1, "Metallography"), 2.8 (in Chapter 2 "Ceramography"), and 3.2 (in Chapter 3, "Plastography"). The marks and descriptions have been taken from the literature and contain important information on hazards and properties.

| Description | Formula | Remarks |
|---|---|---|
| Acetic acid ≤96% | $CH_3COOH$ | C; F; L: colorless aqueous solution, pungent smell, reacts violently with alkaline solutions; H+; D = 1.06 |
| Acetone | $CO\text{-}(CH_3)_2$ | F+ (<21 °C); Xi; L:colorless; D = 0.79 |
| Aluminum chloride | $AlCl_3$ | C; Cr: colorless to pale yellow; H+: fuming when exposed to humid air forming HCl; D = 2.46; violent to explosionlike reaction with water and alcohol |
| Ammonia | $NH_3$ | T; C; Xi; G: becomes liquid in pressure vessels, pungent smell; D = 0.235; explosive with oxidizing gases |
| Ammonium acetate | $CH_3COONH_4$ | Cr: white; H; D = 1.171 |
| Ammonium chloride (salmiac) | $NH_4Cl$ | Xn; Xi; Cr: white, stable in air, salty/bitter taste; D = 1.52 |
| Ammonium citrate | $C_6H_{14}N_2O_7$<br>$(NH_4)_2C_6H_6O_7$ | Cr: white; D = 1.48 |
| Ammonium ferrous sulfate (Mohr's salt) | $(NH_4)_2Fe(SO_4)_2{\cdot}6H_2O$ | Cr: slightly bluish green, sensitive to light; D = 1.86 |
| Ammonium hydrogen fluoride | $(NH_4)HF_2$ | T+; C; Cr: colorless; forms HF when in contact with acids; D = 1.21 |
| Ammonium hydroxide | $NH_4OH$ | C (>35%); Xi; L: aqueous solution; pungent smell; D = 0.882–0.96 |
| Ammonium molybdate | $(NH_4)_6Mo_7O_{24}{\cdot}4H_2O$ | Cr: colorless; releases $NH_3$ when exposed to air |
| Ammonium persulfate | $(NH_4)_2S_2O_8$ | Xi; O; Cr: white, store in dry area, dissociates into $O_2$ and ozone by contact with water; the same reaction occurs when heated; D = 1.98; strong oxidizing reagent |
| Ammonium polysulfide | $(NH_4)_2S_x$ | C (>5%); L: yellow |
| Ammonium tartrate | $(NH_4)_2C_4H_4O_6$ | Cr: colorless; releases $NH_3$ when exposed to air |
| Ammonium tetrachlorocuprate II | $(NH_4)_2CuCl_4{\cdot}2H_2O$ | Cr: yellow; H; D = 2.0 |
| Ammonium thiocyanate | $(NH_4)SCN$ | Xn; contact with acids will generate HSCN; Cr: clear; D = 1.3 |
| Ammonium thiosulfate | $(NH_4)_2S_2O_3$ | Cr |
| Antimony trioxide | $Sb_2O_3$ | T; C: white; D = 5.2–5.8 |
| Aqua regia (nitrohydrochloric acid, Kingswater) | 3 vol% HCl<br>1 vol% $HNO_3$ | T; C: generates chlor- and nitrosyl chlorident; L: yellow with pungent odor, do not store; extremely corrosive |

(continued)

# List of Chemicals (continued)

| Description | Formula | Remarks |
|---|---|---|
| Argon | Ar | G (liquid below −189.2 °C); D = 0.536 |
| Benzene | $C_6H_6$ | F+; T; Xi; I (<21 °C); L: colorless, aromatic smell; D = 0.88; is explosive with air and strong oxidizing agent |
| Benzoyl peroxide | $C_{14}H_{10}O_4$ | E; Xi; Cr: white; organic oxidizing agent |
| Bromine | $Br_2$ | T+; C; L: deep red brown, fuming, pungent smell; D = 3.14; strong oxidizing agent |
| Butanol | $CH_3(CH_2)_3OH$ | F (21–55 °C); Xn; Xi; L: colorless; D = 0.81 |
| 2-butoxyethanol | $C_6H_{14}O_2$ | Xn; Xi; L (not volatile): difficult to ignite; forms peroxides when exposed to air; reacts violently with strong oxidizing agents; D = 0.9 |
| Cadmium chloride | $CdCl_2$ | T; Cr: colorless; mother-of-pearl appearance; H: decomposes by moisture absorption from air to a white powder; D = 4.05 |
| Carbon tetrachloride | $CCl_4$ | T+; L: colorless, very volatile, sweetish aromatic to pungent odor; D = 1.59 |
| Cerium (IV) nitrate | $Ce(NO_3)_4$ | Cr |
| Chromic acid | $H_2CrO_4$ | T; C; L: aqueous solution of chromium trioxide, yellow-red, sour taste; causes fire with organic substances; strong oxidizing agent |
| Chromium (III) oxide | $Cr_2O_3$ | Cr: green; D = 5.22 |
| Chromium (VI) oxide | $CrO_3$ | T; C; Cr: dark red, strong sour taste; H: becomes liquid in air; D = 2.7; is ignitable with flammable substances: strong oxidizing agent |
| Chromosulfuric acid | $H_2SO_4 + CrO_3$ | T+; C; O; L: deep red brown, oil-like consistency; D = ~ 1.85; flammable with ignitable materials; strong oxidizing agent |
| Citric acid | $C_6H_8O_7$ | Cr: white; D = 1.542 |
| Copper ammonium persulfate | $\{Cu(NH_3)\}S_2O_8$ | Cr |
| Copper (II) chloride | $CuCl_2{\cdot}2H_2O$ | Cr: sky blue (green when contaminated); D = 2.51 |
| Copper (II) nitrate | $Cu(NO_3)_2{\cdot}3H_2O$ | Cr: deep blue; stable in air; D = 2.32; strong oxidizing agent |
| Copper (II) sulfate | $CuSO_4{\cdot}5H_2O$ | Cr: glacier blue; D = 2.284 |
| Cryolite | $Na_3AlF_6$ | Xn; Xi; Cr: bright white or transparent; D = 2.95 |
| 1.3-dimethyl, 2-thiourea | $C_3H_8N_2S(CH_3NHCSNHCH_3)$ | Cr; H |
| Ethanol | $C_2H_5OH$ | F+ (<21 °C): produces violent reactions with strong oxidizing agents; can self-ignite; L: colorless, volatile; H; D = 0.79–0.81 |
| Ethylenediaminetetra acetic acid: EDTA | $C_{10}H_{16}N_2O_8$ | Cr: colorless |

(continued)

# List of Chemicals (continued)

| Description | Formula | Remarks |
|---|---|---|
| Ethylene glycol, glycol | $OH\text{-}(CH_2)_2\text{-}OH$ | Xn; Xi; L: colorless, not very volatile, sweetish taste; H; D = 1.11; produces violent reactions and may self-ignite with strong oxidizing agents |
| Fluoroboric acid | $HBF_4$ | C; L: aqueous solution, colorless; D = 1.23 |
| Formic acid | HCOOH | C; L: aqueous solution, colorless and volatile, pungent smell; D = 1.19–1.22 |
| Glacial acetic acid ≥96% | $CH_3COOH$ | C; F (21–55°C); L: glacial acetic acid without water is clear, without color and has a pungent smell; H+; D = 1.05; violent to explosive reactions with alkaline solutions and strong oxidizing agents |
| Glycerol, glycerine | $C_3H_8O_3(HOCH_2CHOHCH_2OH)$ | L: viscous, colorless, clear, sweetish taste; H; D = 1.26; flammable at >100 °C; reacts violently with oxidizing agents, may self-ignite |
| Gold (III) chloride | $AuCl_3$ | Cr; D = 3.9 |
| Hexamethylenetetramine | $C_6H_{12}N_4$ | F; Xi; Cr: white; H |
| Hydrogen | $H_2$ | F+; G (is liquid below −253 °C); D = 0.03; forms explosive mixtures with air and strong oxidizing agents |
| Hydrogen peroxide | $H_2O_2$ | C; O; L: colorless; D = 1.12–1.13 |
| Hydrochloric acid | HCl | C (>25%); Xi; L: clear to yellowish, fuming; D = 1.12–1.19 |
| Hydrofluoric acid | HF | T+; C; L: waterlike solution; pungent smell; H; D = 1.01–1.16 |
| Iodine | $I_2$ | Xn; Cr (also in alcoholic mixture): gray to black, volatile at RT; vapor (violet); D = 4.93; ammonia solution and iodine form explosive iodine nitride |
| Iron (III) chloride | $FeCl_3 \cdot 6H_2O$ | C; Xn; Cr: dirty yellow; H: becomes liquid when exposed to humid air, smells slightly of HCl; D = 2.8 |
| Iron (III) nitrate | $Fe(NO_3)_3 \cdot 9H_2O$ | Cr: nearly colorless |
| Iron (II) sulfate | $FeSO_4$ | Xn; Xi; Cr: with soluble water: light green; without soluble water: white; D = 1.9–2.97 |
| Lactic acid | $C_3H_6O_3$ | L: viscous, colorless; H; D = 1.21 |
| Lead acetate | $Pb(CH_3COO)_2 \cdot 3H_2O$ | T; Xn; Cr: white; D = 2.55 |
| Lithium hydroxide | LiOH | C; Xn; Cr: white; H; D = 1.4 |
| Magnesium oxide | MgO | Cr: white; D = 3.58 |
| Magnesium perchlorate | $Mg(ClO_4)_2$ | Cr: colorless; H; D = 2.21 |
| Mercury (II) nitrate | $Hg(NO_3)_2 \cdot H_2O$ | T; Cr: colorless; H; D = 4.3 |
| Methanol | $CH_3OH$ | T; F+ (<21 °C); L: volatile, colorless; D = 0.79 |
| Methyl methacrylate | $C_5H_8O_2$ | F+ (<21 °C); Xi; L: colorless, clear; D = 0.94; explosive when exposed to air and reacts violently or ignites when exposed to strong oxidizing agents |

(continued)

# List of Chemicals (continued)

| Description | Formula | Remarks |
|---|---|---|
| Molybdic acid | $MoO_3$ | Cr: white, contains ammonium molybdate; D = 4.696 |
| Nitric acid | $HNO_3$ | C; O; L: colorless, fumes when exposed to humid air, pungent smell; D = 1.37–1.52; develops nitrouse gases when concentrated acid is in contact with metals and organic substances |
| Nitrogen | $N_2$ | Xn; G (is liquid below –210 °C); D = 0.31 |
| Orthophosphoric acid | $H_3PO_4$ | C; L: aqueous solution, colorless, viscous; H; D = 1.77–1.88 |
| Oxalic acid | $(COOH)_2$ | Xn; Xi; Cr: colorless; D = 1.9; explosive when oxidizing agents are present |
| Perchloric acid | $HClO_4$ | C, O (>50%); L: colorless, fumes when exposed to air; H+; D = 1.15–1.76; explosive when in contact with oxidizing inorganic and organic substances; strong oxidizing agent |
| Picric acid | $C_6H_3N_3O_7$ | T; E; Cr: yellow, bitter taste; D = 1.8 |
| Potassium carbonate | $K_2CO_3$ | Xi; Cr: white, $CO_2$ will form with acid; H; D = 2.43 |
| Potassium chlorate | $KClO_3$ | Xn; O; Cr: white, D = 2.32; explosive with organic materials; strong oxidizing agent |
| Potassium chloride | $KCl$ | Cr: colorless, salty; D = 1.984 |
| Potassium chromate | $K_2CrO_4$ | Xi; O; Cr: bright yellow; D = 2.73; oxidizing agent; inflammable when in contact with flammable materials |
| Potassium cyanide | $KCN$ | T; Cr: granular; H; D = 1.52; HCN will form when in contact with moist air, carbonized water and acids |
| Potassium dichromate | $K_2Cr_2O_7$ | T; Xi; O: when in contact with flammable materials; Cr: orange red; D = 2.69; strong oxidizing agent |
| Potassium disulfate | $K_2S_2O_5$ | Cr: white; D = 2.34 |
| Potassium ferricyanide | $K_3\{Fe(CN)_6\}$ | Xn; Xi; Cr: intense dark red; D = 1.85; with strong acids HCN will be release; strong oxidizing agent |
| Potassium ferrocyanide | $K_4\{Fe(CN)_6\}\cdot 3H_2O$ | Xn; Xi; Cr: light yellow, stable in air, but decomposes in light; D = 1.853; when in contact with strong acids, HCN will be released |
| Potassium hydrogen carbonate | $KHCO_3$ | Xn; Cr: colorless, transparent or white; D = 2.17 |
| Potassium hydrogen fluoride | $KHF_2$ | T; C; Cr: white; H; D = 2.37; will release HF when heated |
| Potassium hydrogen sulfate | $KHSO_4$ | C; Xi; Cr: white, transparent; D = 2.322 |

(continued)

## List of Chemicals (continued)

| Description | Formula | Remarks |
|---|---|---|
| Potassium hydroxide | KOH | C; Cr; white pellets; H: becomes liquid when exposed to air, absorbing moisture and carbon dioxide; violent exothermic reaction with strong acids; D = 2.04 |
| Potassium hydroxide solution | KOH | C (>5%); L: aqueous solution, colorless, viscous to thin liquid; violent exothermic reaction with strong acids; D = 1.007–1.51 |
| Potassium iodide | KI | C; Xn; Cr: aqueous solution, colorless, decomposes slowly and turns yellow when exposed to air and light; D = 3.12 |
| Potassium nitrate | $KNO_3$ | O; Cr: colorless or white, stable in air, will generate nitrouse gases when heated; D = 2.09; strong oxidizing agent |
| Potassium permanganate | $KMnO_4$ | Xn; O; Cr: deep red, stable in air; D = 2.7; explosive with organic materials and with reducing agents, may ignite; strong oxidizing agent |
| Potassium phthalate | $C_8H_4K_2O_4$ | Cr |
| Potassium thiocyanate | KSCN | Xn; Cr: colorless; H: forms HSCN or HCN when exposed to acids; D = 1.886 |
| Selenic acid | $H_2SeO_4$ | T; L: colorless; H; D = 3.0; strong oxidizing agent |
| Silver cyanide | AgCN | T; Xi; Cr: white, can be exposed to air; D = 3.95 generates cyanides when in contact with strong acids |
| Silver nitrate | $AgNO_3$ | C; O; Cr: colorless, transparent; very sensitive to light; D = 4.35; generates nitrouse gases when heated |
| Sodium carbonate | $Na_2CO_3$ | Xi; Cr; white granular; D = 2.53 |
| Sodium chlorate | $NaClO_3$ | Xn; O; Cr: white; H; D = 2.49; explosive when mixed with organic materials; strong oxidizing agent |
| Sodium chloride | NaCl | Cr: colorless, salty; D = 2.164 |
| Sodium chromate | $Na_2CrO_4 \cdot 10H_2O$ | T; Xi; Cr: yellow; H; D = 1.48; ignites when in contact with flammable material; oxidizing agent |
| Sodium cyanide | NaCN | Xn; Cr; D = 1.86; generates extremely poisonous HCN when exposed to humid air, carbon dioxide containing water, and acids |
| Sodium dichromate | $Na_2Cr_2O_7$ | Xn; Xi; Cr: orange red; H; D = 2.52; ignites when in contact with flammable material; oxidizing agent |
| Sodium fluoride | NaF | T; Cr: white or greenish; H; D = 2.79; generates HF when in contact with acids |
| Sodium hydrogen carbonate | $NaHCO_3$ | Cr: white; stable in air; D = 2.159 |
| Sodium hydrogen phosphate | $Na_2HPO_4 \cdot 12H_2O$ | Xn; Cr; H |

(continued)

# List of Chemicals (continued)

| Description | Formula | Remarks |
|---|---|---|
| Sodium hydroxide | NaOH | C; Cr: white; H: becomes liquid in air by absorbing moisture and carbon dioxide; violent exothermic reaction with strong acids; D = 2.13 |
| Sodium hydroxide solution | NaOH | C (>5%); Xi; L: aqueous solution, colorless, viscous to thin liquid; violent exothermic reactions when in contact with strong acids; D = 1.01–1.53 |
| Sodium hypochlorite | NaOCl | C (>10% chlor); L: aqueous solution with >10% active chlor, light yellow greenish; D = 1.22; generates poisonous chlor gas, with acids; oxidizing agent |
| Sodium peroxide | $Na_2O_2$ | C; O: possible ignition by strong oxidizing effect; Cr: white to yellow; H; store in a dry and closed container, otherwise will decompose by generating $O_2$ and $H_2O_2$ and heat; D = 2.81; fire hazard with flammable substances |
| Sodium sulfate | $Na_2SO_4 \cdot 10H_2O$ | Cr: colorless, decomposes in air; D = 1.464 |
| Sodium sulfide | $Na_2S$ | C; Cr: colorless or yellow to brownish red, reacts with acid generating $H_2S$; D = 1.86 |
| Sodium tetraborate | $Na_2B_4O_7 \cdot 10H_2O$ | Cr: white; D = 1.72 |
| Sodium thiocyanate | NaSCN | Xn; Cr: colorless; H; generates HSCN in presence of acids |
| Sodium thiosulfate; photographic fixer | $Na_2S_2O_3 \cdot 5H_2O$ | Cr: colorless, salty to bitter taste; D = 1.729 |
| Sulfuric acid | $H_2SO_4$ | C (>15%); Xi; L: clear oily consistency, H+; D = 1.18–1.84 |
| Tartaric acid | $C_4H_6O_6$ | Cr: colorless; D = 1.7598–1.788 |
| Tetrafluoromethane | $CF_4$ | G: thermal decomposition generates poisonous and corrosive gases; D = 0.63 |
| Thioglycolic acid | $HSCH_2COOH$ | T; C; L: colorless; strong odor; oily consistency; D = 1.27–1.33 |
| Thiourea | $CS(NH_2)_2$ | Xn; Cr: white shiny, bitter taste; D = 1.405; generates sulfur or $SO_2$ with strong oxidizing agents and when heated |
| Tin (II) chloride | $SnCl_2$ | C; Cr: white; D = 3.95 |
| Toluene | $C_6H_5\text{-}CH_3$ | F; Xn; Xi; L: colorless, volatile, aromatic odor; D = 0.87; explosive with air |
| Trichloroethylene | $C_2HCl_3$ | Xn; Xi; L: colorless, volatile, sweetish etherlike odor; D = 1.462; may be explosive with oxygen |
| Triethylamine | $C_6H_{15}N$ | F+ (<21°C); Xi; L: colorless to yellowish, volatile, oily consistency, has ammonia odor; D = 0.73; vapors form explosive mixtures with air |

(continued)

## List of Chemicals (continued)

| Description | Formula | Remarks |
|---|---|---|
| Vanadium oxide | $V_2O_5$ | Xn; Xi; Cr: rusty red to red; D = 3.36; is a catalyst for many oxidizing reactions, also has an ignition effect; strong oxidizing agent |
| Xylene | $C_6H_4(CH_3)_2$ | F+ (21–55 °C); Xn; Xi; L: colorless to yellowish, volatile, aromatic odor, violent reactions or ignitions with strong oxidizing agents ; vapors form explosive mixtures with air; D = 0.86–0.88 |
| Zinc chloride | $ZnCl_2$ | C; Cr: white; H+; D = 2.91 |

T+, acute poison; T, poison; C, corrosive; E, explosive; F+, highly flammable; F, slightly flammable; Xn, light poison; Xi, irritating; O, supports burning; Cr, crystalline; L, liquid; G, gaseous; H+, very hygroscopic; H, hygroscopic; D, density (g/cm$^3$)

## 4.3 Hazardous Chemical Symbols and Their Interpretation

The pictographs of the hazardous symbols identify the initial information of handling hazardous materials.

| | | |
|---|---|---|
| Toxic | T+/T  | Acute poison and poisons are those substances and preparations that may cause serious health damage or death when ingested or absorbed through the skin or inhaled, use safety equipment. |
| Light poison | Xn  | Substances and preparations that can affect the health to a lesser degree when inhaled, ingested, or absorbed by the skin |

(continued)

## Hazardous Chemical Symbols and Their Interpretation (continued)

| | | |
|---|---|---|
| Corrosive | C  | Substances and preparations that can by contact destroy life tissue. Protective clothing must be worn; use safety goggles and safety gloves. |
| Irritating | Xi  | Substances and preparations that may not cause immediate damage when in contact with skin or mucous tissue, but can cause sensitization of a person to further exposure; use safety goggles and safety gloves. |
| Highly flammable | F+  | Flammable or combustible substances are liquids that have a flash point below 0 °C and a boiling point below 35 °C. They should therefore be removed from open flames, sparks, and heat sources. |
| Light flammable | F  | Substances such as yellow phosphorus (burns by itself), solid materials (e.g., celluloid, which may easily ignite), liquids with flash point below 21 °C (e.g., benzene and acetone), flammable gases such as hydrogen or propane, all substances that generate a flammable gas by contact with water (e.g., sodium). Keep away from flames, sparks, or heat. |
| Oxidizing materials | O  | Substances that support the burning of flammable materials |

(continued)

## Hazardous Chemical Symbols and Their Interpretation (continued)

| | | |
|---|---|---|
| Explosives | E  | Explosive materials and substances may cause explosion under certain conditions. Impact, friction, sparks, and heat exposure must be avoided. |

## 4.4 Handling Tips for Hazardous Materials

Working with chemicals or hazardous materials always requires special care and precaution. Even at low concentration they are often harmful to the human organism, and their absorption into the human body often goes unnoticed. Law and labor associations have provided numerous guidelines that must be observed when handling hazardous materials. Some rules, tips, and precautions are listed below, but naturally they do not cover all the hazardous materials.

### *4.4.1 Rules, Tips, and Precautions*

- When using hazardous materials, legal guidelines applicable for the individual country must be followed.
- Smoking is not permitted in the laboratory. Eating and drinking should also be prohibited.
- All chemicals must be identified with labels that are easily legible.
- When *mixing the etchant,* always add the aggressive chemicals into the diluents (water, alcohol, glycerol, etc.).
- When *working with corrosive chemicals,* always use protective gear such as safety glasses (face shields), protective gloves, and apron. If possible, work with a fumehood.
- *Hazardous materials* should not be discarded in the sink, but should be collected and properly disposed of according to local regulations. This applies especially to *poisonous* and *extremely poisonous* materials.
- *Explosive mixtures,* such as mixtures of nitric acid and glycerol (nitroglycerol), should only be mixed for immediate use. Neutralize any remnants directly, for example, with soda ($Na_2CO_3$).

- When working with *flammable materials,* suitable extinguishers should be at hand (e.g., D-type extinguishers for metal fires).
- If chemicals come into contact with *eyes,* thorough rinsing with water (at least 15 min) is necessary. Eye wash stations are obligatory.
- *Containers with hazardous materials* that require handling with both hands must be stored on low shelves and cupboards (<170 to 175 cm).
- *Breakable containers* should not be carried at their bottle neck, but must be supported at the bottom. A safety container must be used when carrying them over longer distances.
- *Corrosive, aggressive, or slightly poisonous chemicals* must be stored safely away when not in use.
- *Explosive chemicals* must be stored separately from other chemicals.
- The amount of *hazardous materials* in the laboratory should never exceed the amount for immediate use (approximately 1 L at a time).
- All *hazardous materials* and their *preparations* should be examined at least once a year for their appropriate condition; if necessary they should be rebottled or discarded.
- *Fuming chemicals,* such as ammonium hydroxide, hydrofluoric acid, nitric acid, and hydrochloric acid, must be stored in well-ventilated locations.

### 4.4.2 Important Special Precautions

The etching recipes (see sections 1.9, 2.8, and 3.3) and the list of chemicals (see section 4.2 in this Appendix) mention where especially hazardous materials are used. Specific hazards that are frequently encountered in metallography and therefore need special precautions are listed below:

**Perchloric acid** in concentrations exceeding 60% is highly flammable and explosive. This danger is greatly increased by the presence of organic materials or by metals such as antimony, bismuth, and steel, which oxidize readily. Higher concentrations and heating of these solutions should be avoided, particularly in electrolytic polishing and etching. Never store highly concentrated perchloric solutions in plastic containers. When mixing perchloric acid and alcohol, highly explosive alkyl perchlorates may form. High concentrations and high temperatures must also be avoided in those cases.

All *perchloric-acid-containing mixtures* are basically flammable and explosive. Always add perchloric acid slowly and under constant stirring. The temperature of the solution during the mixing must never exceed 35 °C (95 °F); a cooling bath should therefore be used. If possible, work under a special perchloric acid fume hood.

The tendency of perchloric acid to explode is directly related to the concentration; this is illustrated in Fig. 4.1 in the ternary diagram for perchloric acid, water, and acetic acid. However, even compositions outside the critical range can become dangerous with an increase in localized concentration during polishing or etching (water evaporation, temperature increase by contact with organic matters and also with inorganic materials, such as bismuth, or by sparks). Extreme caution should therefore always be exercised when working with perchloric acid, and sufficient stirring and cooling of the electrolyte must be ensured.

Commercially available perchloric acid electrolytes and those polishing and etching recipes listed in this book are not dangerous when handled correctly and appropriately. Their individual components are separately packed during shipment. The perchloric acid is always added before use to the solvent; never the reverse!!!

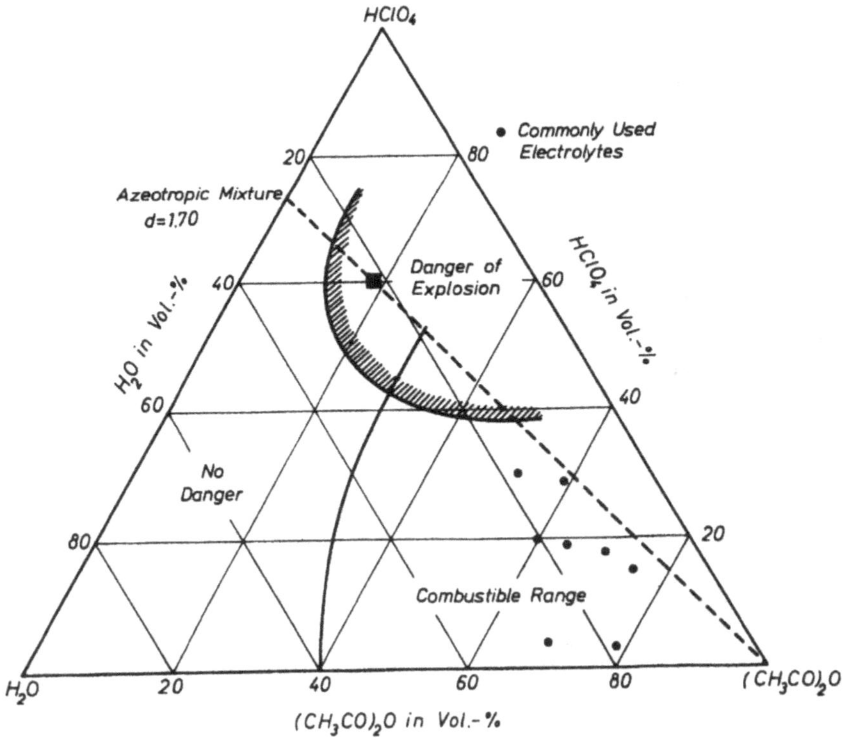


**Fig. 4.1** Explosive, flammable, and safe compounds of perchloric acid, acetic acid hydride, and water mixtures

Spilled perchloric acid must be heavily diluted with water. Do not wipe with cotton or cellulose.

**Mixtures of alcohol with an acid,** for example, perchloric acid, hydrochloric acid, sulfuric acid, nitric acid, or phosphoric acid, can form esters. Some esters are potent nerve poisons that can be absorbed through the skin. They can also cause great damage when inhaled.

**Mixtures of alcohols with nitric acid** can form different reaction products (aldehydes, carbonic acids, explosive nitrogen compounds, etc.). The tendency to explode increases with decreasing molecular size of the produced component. A nitric acid content of 5% should not be exceeded with ethanol and 35% with methanol. Do not store these mixtures.

**Mixtures of methanol and sulfuric acid** can form dimethyl sulfate; it is odorless and tasteless, but extremely poisonous. When sufficient amounts are absorbed through the skin, it can be fatal (even a gas mask does not offer protection). On the other hand, sulfates of the higher alcohols are not dangerous poisons.

**Concentrated mixtures of ethanol and hydrogen peroxide** can produce dangerous chemical reactions where heat and $CO_2$ may be generated by explosion. *Do not* store these mixtures!

**Mixtures of chromium (VI) oxide with organic materials** (e.g., alcohol) are explosive. Mix carefully! Do not store! Never discard them in a sink!

**All cyanide mixtures** are extremely dangerous. If combined with acids, they may form *hydrocyanic acid (HCN)*. This can be lethal even at low concentrations (as vapor in the air, it smells of bitter almonds). Therefore: *do not use with acids! Lethal!*

Use a separate fumehood when working with KCN (potassium cyanide) and store used solution in dedicated bottles.

**Hydrofluoric acid** aggressively attacks the skin. Use protective cream for hands and wear safety gloves! Hydrofluoric acid is also a respiratory poison. Always work under a fume hood. Hydrofluoric acid attacks glass, and it must therefore be stored in plastic containers! After etching with hydrofluoric- acid-containing solutions and before examinations under the microscope, the samples must be very thoroughly rinsed (at least 20 min) and dried. Any adhering hydrofluoric acid will damage the outside lens of the objective lens.

**Hydrogen peroxide and potassium hydroxide** must be mixed very carefully; this mixture produces a high gas evolution that will increase with extra heat and the peroxide will break down. Do *not* store the mixture!

**Aqua Regia.** When aqua regia turns dark brown, discard or poisonous nitrosyl chloride will form. *Never* store in gastight bottles.

**Molten Salts.** Never add water to molten salts (danger of splashing!). Also, *never* add fresh salt to molten salts since it may contain moisture. When working with molten NaOH use only Ni or Fe crucibles; platinum crucibles will be attacked.

**Toxic Materials.** When preparing metallographic samples of toxic materials, for example, thallium, beryllium, or radioactive materials (uranium, plutonium, and their alloys), a glove compartment (closed systems) or "hot cells" must be used.

**Highly inflammable dust or chips** as they are generated when working metals (e.g., magnesium, uranium, zirconium, etc.) are easily ignitable. Any work must be carried out under a protective environment such as water, nitrogen, argon, and so forth. Inflammable substances must *not* be stored in their vicinity; if possible work in a tiled room!

## 4.4.3 Dangerous Reactions of Chemicals

The efficiency of the etching recipes depends on the chemicals used and their mixing ratios. In most cases, the recipes have been established empirically. Even when modifying some proven recipes or when developing new ones, a "quick try" is usually preferred to a more analytical approach. However, this can lead to unexpected and dangerous reactions of the individual mixing components. The following list shows the most common reactions that may occur during experimenting.

| Chemical | Dangerous reaction with: |
|---|---|
| Acetone | Chloroform, chromium (VI) oxide, hydrogen peroxide |
| Alcohols | Perchloric acid, mercury nitrate, hydrogen peroxide, chromium (VI) oxide |
| Aluminum chloride | Water, alcohols |
| Ammonium hydroxide | Metal powders, acids, mercury, hydrofluoric acid |
| Bromine, iodine | Ammonium hydroxide, unsaturated reagents, metal powders, alkaline, and rare earth metals, hydrocarbons, and light (use dark-colored bottles) |
| Chlor | Aluminum and aluminum salts, acids, metal powders, sulfuric acid, cyanides, finely dispersed organic substances, inflammable materials; see also perchloric acid |
| Chromium (VI) oxide | Ammonium hydroxide, glacial acetic acid, glacial acetic acid anhydrous, glycerol, alcohol, organic materials, sodium, inflammable liquids |
| Cyanides | Acids, oxidizing agents, especially nitrites and mercury (II) nitrate |
| Formic acid | Aluminum, oxidizing agents |
| Glacial acetic acid | Chromium (VI) oxide, nitric acid, perchloric acid, hydrogen peroxide, permanganate, alcohol, ethylene glycol, potassium hydroxide |
| Halogen hydrocarbons | Alkaline and earth alkaline metals, concentrated bases, metal powders, perchloric acid, nitric acid, aluminum containers |

(continued)

## *Dangerous Reactions of Chemicals (continued)*

| Chemical | Dangerous reaction with: |
|---|---|
| Hydrofluoric acid | Ammonium hydroxide, alkali metals, glass |
| Hydrochloric acid | Aluminum, organic substances |
| Hydrogen peroxide | Copper, chromium, iron, metals and their salts, alcohol, acetone, organic substances or other flammable materials, formic acid, bronze, glycerol, potassium or sodium hydroxide, potassium permanganate |
| Nitric acid | Ammonium hydroxide, acetic acid anhydrous, chromium (VI) oxide, hydrocyanic acid, flammable liquids and gases, ethanol |
| Oxalic acid | Silver, mercury, oxidizing agents |
| Perchloric acid | Acetic acid, acetic acid anhydrous, bismuth and bismuth alloys, alcohol, concentrated sulfuric acid, concentrated phosphoric acid, substances that oxidize, paper, and wood |
| Peroxide | Acetone, alkali hydroxide, alcohol, formic acid, glycerol, potassium or sodium hydroxide, potassium permanganate, heavy metals (including oxides and salts), dust, oxidizing materials, ammonium hydroxide |
| Picric acid and picrates | Heat, ammonium hydroxide, alkaline and earth alkaline metals, aluminum, nitric acid, peroxide, oxidizing media, sulfuric acid |
| Potassium hydroxide | Glacial acetic acid, aluminum, zinc, halogen hydrocarbons, water |
| Potassium permanganate | Oxidizing organic and inorganic materials, hydrochloric acid, concentrated mineral acids, sulfuric acid, hydrogen peroxide, glycerol, ethylene glycol |
| Silver nitrate | Ammonium hydroxide, ethanol, sodium hydroxide |
| Sodium hydroxide | Aluminum, concentrated acid, silver nitrate, trichloroethylene, chloroform, water, hydrogen peroxide |
| Sodium peroxide | Aluminum, aluminum persulfate, magnesium, organic materials, methanol, ethanol, glycerol, ethylene glycol, glacial acetic acid, acetic acid anhydrous |
| Sulfuric acid | Potassium (per)chlorate, potassium permanganate, flammable substances, sodium carbonate, sodium hydroxide, picrates, water |

## *4.4.4 Disposal of Hazardous Materials*

Legal rules and regulations must be observed when disposing chemicals.

- As a rule, hazardous materials should be collected and disposed of by an authorized company.
- Chemicals should either be neutralized, oxidized, or reduced!
- Spilled acid should be treated with calcium hydroxide and or sodium hydrogen carbonate powder, then rinsed with water.
- Water-soluble inorganic hydroxides, alkalines, and organic basic solutions should be slowly neutralized with and acid (possibly sulfuric acid) and then drained with an ample amount of water.

- Hydrofluoric waste or organic fluor compounds must be precipitated as calcium fluorides by using calcium hydroxides.

**Solvents.** Three different possibilities exist for the disposal of solvents such as acetone, butanol, butyl cellosolve, ethylene glycol, methanol, toluene, trichloroethylene, and xylene: collection for the disposal by an authorized company, burning at a special facility, and recycling by redistillation. Ethanol can be easily redistilled.

All solvents must be separately collected, according to the following:

- Hydrocarbons that cannot be mixed with water: toluene, xylene
- Hydrocarbons that can be mixed with water: ethanol, methanol, butanol, acetone, butyl cellosolve, ethylene glycol, benzene
- Chlorinated hydrocarbons: carbon tetrachloride, trichloroethylene

# 5 Literature References

The literature references listed below are recommended for further and more in-depth information. Though they represent a selection of supplemental publications, not all have been used as source material for the publication of this book.

## Journals


- *Advanced Materials and Processes,* monthly, English, ASM International, Materials Park, OH 44073
- *Journal of the American Ceramic Society,* monthly, English, American Ceramic Society, P.O. Box 6136, Westerville, OH, 43086-6136
- *Materials Characterization,* monthly, English, Elsevier Science Publishing, 655 Avenue of the Americas, New York, NY 10010
- *Materials Science and Engineering,* monthly, English, Elsevier Sequoia S.A., P.O. Box 851, 1001 Lausanne, Switzerland
- *Materials World,* monthly, English, Institute of Materials, 1 Carlton House Terrace, London, SW1Y 5DB, Great Britain
- *Praktische Metallographie (Practical Metallography),* monthly, German/English, Carl Hanser Verlag, Kolbergstrasse 22, 81679 Munich, Germany
- *Structure,* 2 to 3 times per year, English/German/French, Struers A/S, Valhois Alle 176, DK-2610 Rodovre, Denmark
- *Zeitschrift für Metallkunde (Journal of Materials Research and Advanced Technology),* monthly, German/English, Carl Hanser Verlag Kolbergstrasse 22, 81679 Munich, Germany


## General Textbooks and Reviews


- S. Amelinckx et al., Ed., *Handbook of Microscopy: Application in Materials Science, Solid State Physics and Chemistry,* VCH Verlagsgesellschaft mbH, Weinheim, Germany, 1997

- S. Amelinckx, D. van Dyck, J. van Landuyt, and A. van Tandeloo, *Electron Microscopy: Principles and Fundamentals,* VCH Verlagsgesellschaft mbH, Weinheim, Germany, 1997
- M.F. Ashby, *Materials Selection in Mechanical Design,* Pergamon Press, Oxford, New York, Seoul, Tokyo, 1992
- C. Bagnall and G.F. Vander Voort, Ed., *Failure Analysis: From Large Structures to Small Components,* Elsevier Science Publishing, New York, 1991
- R. Barer and V.E. Cosslett, *Advances in Optical and Electron Microscopy,* Vol 10, Academic Press, London, 1987
- J.K. Beddow, *On Potential Application of Morphological Analysis in Metallography,* Oxford and IBH Publishing Co., New Delhi, India, 1985
- M.E. Blum, *Non-Destructive Metallography and Corrosion,* National Association of Corrosion Engineers, Houston, TX, 1986
- E.L. Brown, T.A. Whipple, and G. Krauss, *Metallography of Duplex Stainless Steel Castings,* American Society for Metals, Metals Park, OH, 1984
- M.G. Burke, E.A. Clark, and E.J. Palmiere, Ed., *Understanding Microstructure: Key to Advances in Materials,* ASM International, Materials Park, OH, 1997
- R.W. Cahn, P. Haasen, and E.J. Kramer, Ed., *Characterization of Materials,*Vol 2, *Materials Science and Technology,* E. Lifshin, Vol Ed., VCH Verlagsgesellschaft mbH, Weinheim, Germany, 1991
- C. Giacovazzo, *Fundamentals of Crystallogaphy,* Oxford Science Publications, Oxford, 1992
- R.C. Gifkins, *Optical Microscopy of Metals,* American Elsevier Publishing, New York, 1970
- T. Gladman, *Quantitative Metallography: Recent Experience with Automatic Image Analysis,* The Institute of Materials, London, 1987
- P.J. Goodhew, *Specimen Preparation for Transmission Electron Microscopy of Materials,* Oxford University Press, Oxford, 1984
- F. Habashi, *Alloys (Preparation, Properties, Applications),* Wiley–VCH, Weinheim, New York, Toronto, 1998
- T.W. Heaslip, *Metallography in Failure Analysis,* Plenum Press, New York, 1978
- E. Hornbogen and G. Petzow, Metallography: State of the Art Survey, *Praktische Metallographie,* Vol 28 (No. 7 and 8), 1991, p 320–332 (No. 7), p 383–403 (No. 8)
- P.J. Kenny, G.S. Cole, D.O. Northwood, J. Wylie, and G.F. Vander Voort, *Image Analysis and Metallography,* ASM International, Materials Park, OH, 1989
- A. Kneissl, F. Jeglitsch, *Fortschitte in der Metallographie (Progress in Metallography),* Vol 30, G. Petzow, Series Ed., Werkstoff-Informationsgesellschaft mbH, Frankfurt, Germany, 1999

- A. Kulmburg, *The Quantitative Metallography of Weld Metal Microstructures,* Pergamon Press Ltd., Oxford, 1981
- M.F.C. Ladd and R.A. Palmer, *Structure Determination by X-Ray Crystallography,* 2nd ed., Plenum Press, New York, 1985
- I. LeMay, *Failure Mechanisms and Metallography: A Review,* Plenum Press, New York, 1978
- I. LeMay and N.O. Beltran, *Metallography and Fractography in Corrosion-Related Failures,* National Association of Corrosion Engineers, Houston, TX, 1986
- D. Loehe and O. Voehringer, Metallic Structural Materials: Design of Microstructure, *Universal Design Theory,* H. Grabowski, S. Rude, and G. Grain, Ed., Shaker Verlag, Aachen, Germany, 1988, p 147–165
- M.R. Louthan, Jr., I. LeMay, and G.F. Vander Voort, *Welding, Failure Analysis, and Metallography,* ASM International, Materials Park, OH, 1998
- J. MacCall and P.M. French, *Metallography as a Quality Control Tool,* Plenum Press, New York and London, 1980
- M.M. McDonald, *The Metallurgy and Metallography of Braze Joints,* ASM International Materials Park, OH, 1987
- G. Petzow, S. Takajo, and W.A. Kaysser, *Application of Quantitative Metallography to the Analysis of Grain Growth during Liquid-Phase Sintering,* ASTM, Philadelphia, PA, 1984
- T.A. Place, J.D. Braun, W.E. White, and G.F. Vander Voort, *Computer-Aided Microscopy and Metallography,* ASM International, Materials Park, OH, 1992
- C. Richardson and B. Ralph, *The Metallography of a High-Performance Superalloy,* Elsevier North-Holland, Inc., New York, 1981
- M. Schreiner, P. Ettmayer, M. Simon, and W. Wruss, *Metallography as a Quality Control in the Production of Hard Metals,* MPR Publishing Services Ltd., Shrewsbury, U.K., 1983
- H. Schumann, *Metallographie,* Deutscher Verlag für Grundstoffindustrie, Leipzig, Germany, 1991 (in German)
- S.A. Shiels, C. Bagnall, R.E. Witkowski, and G.F. Vander Voort, *Corrosion, Failure Analysis, and Metallography,* American Society for Metals, Metals Park, OH, 1986
- C.St. Smith, *A Search for Structure,* The MIT Press, Cambridge, MA, USA, and London, 1981
- G.E. Smith, *Acoustic Metallography: Using More of the Ultrasonic Data,* American Society of Nondestructive Testing, Columbus, OH, 1985, p 1734–1740

- S. Somiya, R.P.H. Chang, M. Doyama, R. Roy, Ed., *Materials,* Science and Engineering Serving Society, Elsevier, Amsterdam, New York, Tokyo, 1998
- D.W. Stevens, E.A. Clark, D.C. Zipperian, and E.D. Albrecht, Ed., *Advances and Application in the Metallography and Characterization of Microelectronic Components,* ASM International, Materials Park, OH, 1996
- A. Tomer, *Structure of Metals through Optical Microscopy,* ASM International, Materials Park, OH, 1991
- G.F. Vander Voort, *Applied Metallography,* Van Nostrand Reinhold, New York, 1986
- E. Weibel, *Stereological Methods,* Academic Press, London, 1980
- D.K. Williams, A.R. Pellen, and R. Gronsky, *Images of Materials,* Oxford University Press, New York and Oxford, 1991


## Handbooks, Compilations, and Tables


- *Atlas of Microstructures of Industrial Alloys,* Vol 7, *Metals Handbook,* American Society for Metals, Metals Park, OH, 1972
- R. Brook, Ed., *Concise Encyclopedia of Advanced Ceramic Materials,* Pergamon Press, Oxford, New York, Frankfurt, 1991
- H.E. Bühler and H.P. Hougardy, *Atlas of Interference Layer Metallography,* Deutsche Gesellschaft für Materialkunde, Frankfurt, Germany, 1980
- P.F. Chaveriat, G.S. Kim, S. Shah, and J.E. Indacochea, *Low Carbon Steel Weld Metal Microstructures (The Role of Oxygen and Manganese),* ASM International, Materials Park, OH, 1987
- T.B. Massalski, *Binary Alloy Phase Diagrams,* Vol 1 to 3, ASM International, Materials Park, OH, 1990
- *Mechanical Testing,* Vol 8, 9th ed., *Metals Handbook,* American Society for Metals, Metals Park, OH, 1985
- *Metallography and Microstructures,* Vol 9, *ASM Handbook,* ASM International, Materials Park, OH, 1985
- Metals-Mechanical Testing, E 340; Elevated and Low-Temperature Tests, E 381; Metallography, E 407, *Annual Book of ASTM Standards,* Vol 03.01, ASTM, Philadelphia, PA, 1993
- G. Petzow and G. Effenberg, Ed., *Ternary Alloys,* Vol 1–9, Materials Science International Services (MSI), 70569 Stuttgart and VCH Verlagsgesellschaft mbH, Weinheim, Germany, 1994
- J.C. Russ, *The Image Processing Handbook,* CRC Press, Boca Raton, Ann Arbor, London, Tokyo, 1992

- M. Schwartz, *Handbook of Structural Ceramics,* McGraw-Hill, New York, 1992


## Metallographic Preparation


- D.E. Alman, J.A. Hawk, and J.H. Simmons, *The Role of Characterization in Understanding Environmental Degradation of Materials,* ASM International, Materials Park, OH 1998
- B. Altshuller and M. Ryvola, *Aluminum Weldment Metallography,* ASM International, Materials Park, OH, 1987
- W.A. Baeslack, *Metallography of Titanium Alloy Weldments,* ASM International, Materials Park, OH, 1987
- E. Beraha and B. Shipgler, *Color Metallography,* American Society for Metals, Metals Park, OH, 1977
- V. Biss and G. Strabel, *Metallographic Study of Platinum Modified Aluminide Coating on IN738 Base Metal,* ASM International, Materials Park, OH, 1987
- L. Bjerregaard, K. Geels, B. Ottesen, and M. Ruckert, *Metallographers Guide,* Struers Tech A.S, Rodovre, Denmark, 1992
- G.A. Blann, Efficient High Volume, *Metallography,* Elsevier Science Publishing, New York, 1983
- M.E. Blum, P.M. French, R.M. Middleton, and G.F. Vander Voort, *Field Metallography, Failure Analysis, and Metallography,* ASM International, Materials Park, OH, 1987
- B. Bousfield, *Surface Preparation and Microscopy of Materials,* John Wiley & Sons, New York, 1992
- B. Bousfield and D.J. Diaz, *New Concepts of Specimen Preparation,* ASM International, Materials Park, OH, 1989
- G.S. Cole, E. Abranovici, and G. Vander Voort, Microstructural Characterization of Lightweight Structural Materials for Transportation, *Materials Characterization,* special issue, Vol 35, Elsevier Science Inc., New York, 1995
- R.S. Crouse, R.J. Gray, and B.C. Leslie, *Application of Color in Metallography and Photography,* Plenum Publishing, New York, 1978
- H. Gahm and F. Jeglitsch, *Color Methods and Their Application in Metallography,* Elsevier North-Holland, Inc., New York, 1981
- R.J. Gray and L.R. Cornwell, *Unusual Techniques and New Applications of Metallography,* Vol II, American Society for Metals, Metals Park, OH, 1986
- W.J. Huppman and K. Dalal, *Powder Metallurgy Metallography,* Verlag Schmid, Freiburg, Germany, 1986

- W.L. Johns and J.H. Doyle, *Metallography of Solid State Bonds,* ASM International, Materials Park, OH, 1987
- T. Malis, *AEM Specimen Preparation (Something Old and Something New),* The Institute of Materials, London, 1988
- W.A. Mannheimer, A Bibliography of Books Relating to Materials Microscopy, *Metallography,* special issue, Vol 22 (No. 2), March 1989
- V.C. Marcotte, M.W. Ricker, and N.G. Koopman, *Metallography of Soldered Joints,* ASM International, Materials Park, OH, 1987
- J.L. McCall, D.L. Olson, and I. LeMay, *Metallography and Interpretation of Weld Microstructures,* ASM International, Materials Park, OH, 1987
- J.L. McCall and W.M. Mueller, Ed., *Microstructural Analysis, Tools and Techniques, Metallographic Specimen Preparation,* Plenum Press, New York, 1974
- *Metallography of Tin and Tin Alloys,* International Tin Research Institute, Greenford, U.K., 1982
- Mounting of Specimens, *Metallography and Microstructures,* Vol 9, *ASM Handbook,* ASM International, Materials Park, OH, 1985
- J.A. Nelson, *New Abrasives for Metallography,* Elsevier Science Publishing, New York, 1983
- L.E. Samuels, *Metallographic Polishing by Mechanical Methods,* 3rd ed., American Society for Metals, Metals Park, OH, 1982
- D. Scott, *Metallography of Ancient Metallic Artifacts,* University of London, Institute of Archaeology, 1987
- Sectioning, *Metallography and Microstructures,* Vol 9, *ASM Handbook,* ASM International, Materials Park, OH, 1985
- Mc.G. Tegart, *The Electrolytic and Chemical Polishing of Metals in Research and Industry,* Pergamon Press, London, 1959
- G.F. Vander Voort, *Metallography, Principles and Practice,* McGraw Hill, New York, 1984
- H. Waschull, *Metallographic Preparation,* Deutscher Verlag für Grundsoffindustrie, Leipzig, Stuttgart, 1993 (in German)


## Metallic Materials


- J.B. Campbell, An Advance in Powder Metallurgy Aluminum Alloy Etchants, *Metallography,* Vol 18, 1985, p 413
- W.R. Graff and D.C. Sargent, A New Grain-Boundary Etchant for Aluminum Alloys, *Metallography,* Vol 14, 1981, p 69

- P.J. Grobner and A. Poznansky, *Metallography of Base Metal and Weld Joints in High Alloy Cast Duplex Stainless Steels,* ASM International, Materials Park, OH, 1987
- I. LeMay and N.O. Beltran, *Metallography and Fractography in Corrosion-Related Failures,* National Association of Corrosion Engineers, Houston, TX, 1988
- S. Liu, Metallography of HSLA Steel Weldments, *Key Engineering Materials,* Vol 1–18, 1992, p 69–70
- C.T. Sims and W.C. Hagel, *The Superalloys,* John Wiley & Sons, New York, 1972


## Ceramic Materials


- G.N. Babini, M. Haviar, and P. Sajgalik, Ed., *Engineering Ceramics '96: Higher Reliability through Processing,* Kluwer Academic Publications, London, 1997
- V. Carle, U. Schäfer, U. Täffner, F. Predel, R. Telle, and G. Petzow, Ceramography of High Performance Ceramics: Description of the Materials, Preparation, Etching Techniques, and Description of the Microstructure, *Praktische Metallographie (Practical Metallography),* Part I, Ceramographic Etching, Vol 28, 1991, p 359–377; Part II, Silicon Carbide, Vol 28, 1991, p 420–434; Part III, Zirconium Oxide, Vol 28, 1991, p 468–483; Part IV, Aluminum Nitride, Vol 28, 1991, p 542–552; Part V, Silicon Nitride, Vol 28, 1991, p 592–610; Part VI, High-Temperature Superconductor $YBa_2Cu_3O_7$, Vol 28, 1991, p 633–648; Part VII, Boron Carbide, Vol 31, 1994, p 218–233; Part VIII, Aluminum Oxide, Vol 32, 1995, p 54–76
- C. Chatfield and H. Norstrom, Plasma Etching of Sialon, *Journal of the American Ceramic Society,* Vol 66, 1983, p C-168
- I.W. Chen, P.F. Becher, M. Mitomo, G. Petzow, and T.S. Yen, *Silicon Nitride Ceramics,* Materials Research Society, Pittsburgh, 1993
- H.J. Cialone, G.W.E. Johnson, M.E. Blum, G.F. Vander Voort, *Metallography of Advanced Materials,* Twentieth Annual Technical Meeting of the International Metallographic Society, ASM International, Materials Park, OH, 1988
- Y.H. Hu, M.P. Harmer, and D.M. Smyth, Solubility of BaO in $BaTiO_3$, *Journal of the American Ceramic Society,* Vol 68, 1985, p 372–376
- N. Ichinose, *Introduction to Fine Ceramics: Applications in Engineering,* John Wiley & Sons Ltd., 1983

- W.A. Kaysser, M. Sprissler, C. Handwerker, and J.E. Blendell, Effect of a Liquid Phase on the Morphology of Grain Growth in Alumina, *Journal of the American Ceramic Society,* Vol 70, 1987, p 339–343
- W.E. Lee and W.M. Rainforth, *Ceramic Microstructures: Property Control by Processing,* Chapman & Hall, London, 1994
- I.M.D. Sachs and J.A. Pask, Sintering of Mullite-Containing Materials, *Journal of the American Ceramic Society,* Vol 65, 1982, p 65–70
- N.T. Saenz, *Ultrathinning Section Techniques for the Characterization of Brittle Materials,* ASM International, Materials Park, OH, 1990
- M. Schwartz, Ed., *Handbook of Structural Ceramics,* McGraw-Hill, Inc., New York, 1992
- D.C. Zipperian, *Microstructural Preparation and Characterization of Metals and Ceramics,* ASM International, Materials Park, OH, 1990
- D.C. Zipperian and St.N. Chanal, Readying Advanced Ceramics for the Microscope, *Advanced Materials and Processes,* Vol 144 (No. 2), 1993, p 15–21


## Polymer Materials


- J.G. Fatou, Optical Microscopy of Fibers, *Applied Fibre Science,* Vol 1, F. Happey, Ed., Academic Press, London, 1978
- V. Jaussent and G.W. Ehrenstein, The Etching of Plastic Weld Seams–A New Method of Preparation, *Praktische Metallographie,* Vol 24 (No. 11), 1997, p 544–554
- E.M. Langer and M. Schwarz, Metallographic Failure Analysis of Fluoropolymer Components, *Praktische Metallographie,* Vol 31 (No. 4), 1994, p 199–208
- R.H. Olley, Selective Etching of Polymeric Materials, *Science Progress,* Vol 70, 1986, p 17–43
- R.deJ. Santos, J.C. Bruno, M.T.M.B Silva, and R.C.R. Nunes, *Polymer Testing,* Vol 12, 1993, p 393
- L.C. Sawyer and D.T. Grubb, *Polymer Microscopy,* Chapman and Hall, London, New York, Tokyo, 1996
- F. Walter, *The Microtome: Guidelines for Preparation and Cutting,* Leica Wetzlar, Germany, 1980 (in German)
- R.J. Young, *Introduction to Polymers,* Chapman and Hall, London, 1981

# Index

## A

**Abrasive cutting**....9(F), 10–11(T)
cutting wheel machines....11
**Abrasive grains**....10–11
**Abrasive grain sizes**....52–53
**Abrasive wheels**....10–11(T)
application....10(T)
**Acetic acid ≤96%,** formula, hazards and properties....190
**Acetone**
dangerous chemical reactions....202
disposal safety and regulations....204
formula, hazards and properties....190
**Acids,** safety precautions with....201
**Acrylics.** *See also* Polymers.
as mounting materials....14(T)
properties....14(T)
**Adhesive tape,** transparent, to preserve surfaces....43
**Adler etchant,** composition and application guidelines....185
**Advanced ceramics.** *See also* Ceramic materials.
base materials....153
definition....153
product forms....153
properties....153
**Alcohol**
dangerous chemical reactions....202
as rinse for drying the specimen surface....38
safety precautions with....201
**Alumina**....22(T)
as abrasive....23, 24
as abrasive for electrolytic lapping....36
abrasive grain sizes....52
effectiveness for polishing....36–37
**Alumina powder**....16
**Aluminum and aluminum alloys**
chemical polishing....58, 60
electrochemical polishing....60–61
electrolytic etching....63
electrolytic lapping....61
electrolytic polishing....58, 59, 60, 64
etching techniques and preparation of samples....57–64
grinding....57, 59
identification....58
lapping....60
macroetching....61
microetching....62–63
physical etching....64
polishing....58, 59, 60, 64
preparation of samples....57–60
rough polishing....58, 59
shock polishing....58, 59
ultramilling....59
**Aluminum chloride**
dangerous chemical reactions....202
formula, hazards and properties....190
**Aluminum foundry alloys**
grinding....58, 59
polishing....58, 59
**Aluminum metal-matrix composites.** *See also* Ceramic materials.
grinding....59
lapping....60
polishing....59, 60

**Aluminum wrought alloys**
grinding ........ 58, 59
polishing ........ 58, 59
**Alundum** ........ 22(T)
**Americium and americium alloys**
electrochemical polishing ........ 133
electrolytic polishing ........ 133
etching techniques and preparation of surfaces ........ 133–135
grinding ........ 133
microetching ........ 134–135
polishing ........ 133
preparation of surfaces ........ 133
radioactivity ........ 133
**Ammonia**
formula, hazards and properties ........ 190
**Ammonium acetate**
formula, hazards and properties ........ 190
**Ammonium chloride**
formula, hazards and properties ........ 190
**Ammonium citrate**
formula, hazards and properties ........ 190
**Ammonium ferrous sulfate**
formula, hazards and properties ........ 190
**Ammonium hydrogen fluoride**
formula, hazards and properties ........ 190
**Ammonium hydroxide**
dangerous chemical reactions ........ 202
formula, hazards and properties ........ 190
**Ammonium molybdate**
formula, hazards and properties ........ 190
**Ammonium persulfate**
formula, hazards and properties ........ 190
**Ammonium polysulfide**
formula, hazards and properties ........ 190
**Ammonium tartrate**
formula, hazards and properties ........ 190
**Ammonium tetrachlorocuprate II**
formula, hazards and properties ........ 190
**Ammonium thiocyanate**
formula, hazards and properties ........ 190
**Ammonium thiosulfate**
formula, hazards and properties ........ 190
**Anodizing** ........ 43
hafnium, pure ........ 112
lanthanum and lanthanum alloys ........ 145
rare earth metals and their alloys ........ 145, 146
refractory metals and their alloys ........ 88
zirconium and zirconium alloys ........ 112
**Anodic reaction** ........ 40
**Antimony and antimony alloys**
etching techniques and preparation of samples ........ 70–72
macroetching ........ 71–72
microetching ........ 72
preparation of samples ........ 70–71
**Antimony trioxide** ........ 36
formula, hazards and properties ........ 190
**Antiphase boundaries** ........ 2
**Aqua regia**
composition and application guidelines ........ 185
formula, hazards and properties ........ 190
safety precautions with ........ 201
**Arc cutting** ........ 9(F)
**Argon,** formula, hazards and properties ........ 191

# B

**Bakelite**
as mounting materials ........ 14(T)
properties ........ 14(T)
**Barker etchant,** composition and application guidelines ........ 185
**Baumann print etchants,** composition and application guidelines ........ 185
**Benzene**
disposal safety and regulations ........ 204
formula, hazards and properties ........ 191
as rinse for drying the specimen surface ........ 38
**Benzoyl peroxide**
formula, hazards and properties ........ 191
**Beraha I etchant,** composition and application guidelines ........ 185
**Beraha II etchant,** composition and application guidelines ........ 186
**Beraha III etchant,** composition and application guidelines ........ 186
**Beraha III 1 etchant,** composition and application guidelines ........ 186

**Beryllium and beryllium alloys**
electrochemical polishing............68–69
electrolytic etching............70
electrolytic lapping............69
electrolytic polishing............67, 68
etching techniques and
preparation of surfaces............67–70
grinding............67, 68, 69
macroetching............69
microetching............69–70
physical etching............70
polishing............67, 68–69
preparation of surfaces............67–68
shock polishing............68
vibratory polishing............68
**Bioden replica technique,** material used, application technique, and microscopic resolution............51(T)
**Bismuth alloys**
embedding of sample in............17
**Bismuth and bismuth alloys**
electrolytic polishing............71
etching techniques and
preparation of samples............70–72
grinding............70, 71
macroetching............71–72
microetching............72
polishing............71
preparation of samples............70–71
vibratory polishing............71
**Bitter technique**............148
composition and application
guidelines............186
**Boride ceramics.** *See also* Ceramic materials.
electrolytic etching............169
microetching............169
**Breakouts**............17
in ceramic materials............156, 157(F), 158–159, 172
in refractory metals............81
**Brightening, anodic**............32
**Brightfield illumination**............39
copper and copper alloys............94
polymers............183
**Bromine**
dangerous chemical reactions............202
formula, hazards and properties............191
**Butanol**
disposal safety and regulations............204
formula, hazards and properties............191
**2-butoxyethanol**
formula, hazards and properties............191
**Butyl cellosolve**
disposal safety and regulations............204

## C

**Cadmium and cadmium alloys**
chemical polishing............73, 74
electrochemical polishing............74
electrolytic etching............75
electrolytic polishing............73, 74
etching techniques and
preparation of samples............72–75
grinding............73
machining............73
microetching............74–75
polishing............73, 74
preparation of samples............72–74
ultramilling............73
**Cadmium chloride**
formula, hazards and properties............191
**Carbide ceramics.** *See also* Ceramic materials.
electrolytic etching............171–172
microetching............169–171
**Carbon tetrachloride**
disposal safety and regulations............204
formula, hazards and properties............191
**Carapella etchant,** composition and application guidelines............186
**Cast iron.** *See also* Iron, steel, and cast iron.
etching techniques and
preparation of surfaces............94–102
preparation of surfaces............94–96
**Cast liquid film technique**
polymers............183
**Cathodic reaction**............40
**Cathodic sputtering**............45
**Cathodic vaporizing**............45

**Cemented carbides.** *See also* Ceramic materials.
electrolytic etching ........................174
microetching ........................173–174
physical etching ........................174
preparation of surfaces ........................172–173
**Ceramic materials** ........................164–167
base materials ........................153
breakouts ......156, 157(F), 158–159, 172
ceramographic etching ........................159
chemical etching ........................160–162
chemical polishing ........................158
cleaning ........................158–159
darkfield illumination ........................160
deformation ........................156
deformation twins ........................156
differential interference contrast ......160
dislocations ........................156
dissolution etching ........................160
electrochemical etching ........................159(F), 160–162
electrolytic etching ..........161, 167, 168, 171–172, 174
electrolytic grinding ........................158
electrolytic polishing ........................158
etching techniques and preparation of surfaces ........................163–174(T)
etch polishing ........................161–162
grinding ........................154, 155–158, 163(T), 164(T), 173
identification ........................155
infiltration ........................155
interference layers ........................162
ion etching ........................162
lapping ........................157, 158, 163(T), 164(T), 173
lubricants for ........................157–158
microetching ........................164–167, 168, 169–171, 172, 173–174
microtome cutting ........................158
mounting ........................154–155
optical etching ........................159–160(F)
physical etching ...159(F), 162–163, 174
plasma etching ........................161
polarization ........................160
polishing ........................155–158(F), 163(T), 164(T), 173
porosity ........................158
preparation of surfaces ........................153–154
reactive sputtering ........................162–163
reflective layer ........................159–160
sampling ........................154
scratches ........................158
sectioning ........................154, 156(F)
smear layers ........................156
thermal etching ........................162
ultramilling ........................158
**Ceramographic examination** ........................3
**Ceramography** ........................153–174(F,T)
definition ........................153
**Cerium and cerium alloys.** *See also* Rare earth metals and their alloys.
etching techniques and preparation of surfaces ..........144–148
flammability ........................144
**Cerium (IV) nitrate,** formula, hazards and properties ........................191
**Cerium-mischmetal, technical grade,** etching techniques and preparation of surfaces. *See also* Rare earth metals and their alloys. .........144–148
**Cerium oxide** ........................22(T)
**Cermets,** preparation of surfaces. *See also* Ceramic materials. ........................154
**Chemicals**
dangerous reactions possible ....202–203
descriptions, formulas, hazards and properties ..........190–196
**Chlor,** dangerous chemical reactions ........................202
**Chlorinated hydrocarbons**
disposal safety and regulations .........204
**Chromic acid**
as chemical polishing solution ............34
formula, hazards and properties........191
**Chromic oxide** ........................22(T)
**Chromium and chromium alloys,** etching techniques and preparation of surfaces. *See also* Refractory metals and their alloys. ........................80–88
**Chromium (III) oxide**
formula, hazards and properties........191

**Chromium (VI) oxide**
dangerous chemical reactions ........... 202
formula, hazards and properties ........ 191
safety precautions with ..................... 201
**Chromosulfuric acid**
composition and application guidelines ...................................... 186
formula, hazards and properties ........ 191
**Citric acid**
formula, hazards and properties ........ 191
**Clamping** ............................... 11, 12–13(F)
examples of fixtures ....................... 12(F)
**Clamps,** materials for ............................ 12
**Cleaning** ................................. 8, 36–37(F)
ceramic materials ...................... 158–159
polymers ............................. 179, 180(T)
**Cleaving** ............................................. 9(F)
**Cobalt and cobalt alloys**
chemical polishing ........................ 77–78
electrochemical polishing ............. 77–78
electrolytic etching ....................... 79–80
electrolytic polishing .............. 75, 76, 77
etching techniques and preparation of samples .............. 75–80
grinding .................................. 75, 76, 77
lapping ................................................ 77
macroetching ....................................... 78
microetching ................................. 78–79
physical etching .................................. 80
polishing ................................. 75, 76, 77
preparation of surfaces .................. 75–77
shock polishing ................................... 76
**Cobalt-base superalloys,** etching techniques and preparation of surfaces ..................... 76, 78, 79, 80
**Cobalt magnetic alloys,** etching techniques and preparation of surfaces ........................... 76, 78, 80
**Coercivity** ............................................... 7
**Coinage silver,** etching techniques and sample preparation. *See also* Silver and silver alloys. .................. 54
**Coolants** ............................................ 9, 11
for ceramic materials ................ 157–158
for grinding ..................................... 25–26
for polishing rare earth metals and their alloys ............................ 146
for polymers ...................................... 177
**Copper ammonium persulfate**
formula, hazards and properties ........ 191
**Copper and copper alloys**
chemical polishing ........................ 90, 91
electrochemical polishing ............ 90–91
electrolytic etching ............................. 94
electrolytic lapping ....................... 90, 91
electrolytic polishing .................... 90–91
etching techniques and preparation of surfaces .............. 88–94
etch polishing ..................................... 89
grinding ............................................... 89
macroetching ....................................... 92
microetching ................................. 92–94
physical etching .................................. 94
polishing ...................................... 89–90
preparation of surfaces .................. 88–90
ultramilling ......................................... 89
**Copper (II) chloride**
formula, hazards and properties ........ 191
**Copper (II) nitrate**
formula, hazards and properties ........ 191
**Copper (II) sulfate**
formula, hazards and properties ........ 191
**Coring (segregation etching)** ................ 42
definition ............................................. 49
**Corrosion resistance** .............................. 7
microstructure-sensitive properties .. 2–3
**Corrosive,** hazardous chemical symbol and its interpretation ........ 197
**Corrosive chemicals,** protective gear worn during work ................. 198
**Corundum**
natural ........................................... 22(T)
precious ......................................... 22(T)
synthetic ........................................ 22(T)
with alumina .................................. 22(T)
**Crowell etchant,** composition and application guidelines .................. 187
**Cryolite**
as coating material for vapor deposition ........................................ 44
formula, hazards and properties ........ 191
**Current-density/voltage curve** .................................. 31–32(F)
**Cutting** ................................................ 9(F)
**Cyanide mixtures**
safety precautions with ..................... 201

**Cyanides**
dangerous chemical reactions...........202
**Czochralski etchant,** composition and application guidelines............187

## D

**Darkfield illumination**
ceramic materials..............................160
copper and copper alloys ....................94
metallographic etching .......................39
polymers...........................................179
**Decarburization** ....................................18
**Deformation**
aluminum and aluminum alloys..........57
cadmium and cadmium alloys ............73
ceramic materials..............................156
cobalt and cobalt alloys.......................75
copper and copper alloys ....................89
indium and indium alloys ...................73
iron, steel, and cast iron ......................94
lead and lead alloys...................125–126
magnesium and magnesium alloys ...114
nickel and nickel alloys.....................119
refractory metals and their alloys .......80
silver and silver alloys ........................53
thallium and thallium alloys ...............73
tin and tin alloys...............................135
titanium and titanium alloys .............139
zinc and zinc alloys...........................148
**Deformation twins**
in ceramic materials..........................156
in hafnium and zirconium .................109
in tin and tin alloys...........................135
in zinc and zinc alloys.......................148
**Dense ceramics and hard metals,** preparation of surfaces. *See also* Ceramic materials. ....163(T)
**Density,** as microstructure-insensitive property .......................................2–3
**Desiccators,** for storage of metallographic samples ..................43
**Detergents**
added to suppress Rehbinder effect ....26
**Diallyl phthalate**
as mounting materials....................14(T)
properties .......................................14(T)
**Diamond**
as abrasive ...................................23, 24
abrasive grain sizes .......................52–53
**Differential interference contrast (DIC)**
ceramic materials..............................160
polymers...........................................179
technique ............................................39
**Diffusion, coefficient of**..........................7
**Diffusion rate,** microstructure-sensitive properties ....................................2–3
**Dimensions,** spectrum of.........................1
**1.3-dimethyl,2-thiourea,** formula, hazards and properties ..................191
**Direct lattice imaging**........................1(F)
**Dislocations,** in ceramic materials .......156
**Dissolution,** degree of ...........................53
**Dix-Keller etchant,** composition and application guidelines..................187
**Ducat gold,** etching techniques and preparation of samples. *See also* Gold and gold alloys. .....................64
**Dysprosium and dysprosium alloys,** etching techniques and preparation of surfaces. *See also* Rare earth metals and their alloys. ..................................144–148

## E

**EDTA (ethylenediaminetetra acetic acid),** formula, hazards and properties...............................191
**Elastic spacers** ......................................13
**Electric resistance** ..................................7
**Electrochemical machining**...............9(F)
**Electrochemical potential** ..............40–41
**Electrocorundum** ............................22(T)
**Electroengraving**.............................18(F)
**Electroless coating** ...............................16
**Electrolytes.** *See also* Etching, electrolytic. ...................................30
**Electrolytic polishers** ...........................30
**Electrolytic sample preparation**....16–17
**Electromotive series of elements**..........41
**Electropolishing**........................30–33(F)

**Electron microscopy,** magnification and resolution ranges ....4(F)
**Electron optical microscopy** ....3, 4
**Elongation** ....7
**Embedding** ....11, 12(F), 13–15(F)
simultaneous ....18(F)
**Emery** ....22(T)
**Engraving** ....18(F)
**Epoxy resins**
as mounting materials ....14(T)
properties ....14(T)
**Erbium and erbium alloys,** etching techniques and preparation of surfaces. *See also* Rare earth metals and their alloys. ....144–148
**Esters,** safety precautions with ....201
**Etchants** ....8, 22, 41–42
designation of ....185–189
for etch polishing ....35
for polymers ....180–181
mixing of ....198
names ....53
**Etching** ....8
**Etching, anodic** ....42–43
definition ....47
**Etching, cathodic ion** ....45, 46(F)
**Etching, ceramographic**
ceramic materials ....159
**Etching, chemical** ....40–43
ceramic materials ....160–162
definition ....48
polymers ....179–180
**Etching, cold** ....47
definition ....47
**Etching, color immersion** ....42
**Etching, controlled,** definition ....47
**Etching, conventional** ....45
**Etching, conventional chemical** ....42
**Etching, crystal figure,** definition ....47
**Etching, deep,** definition ....48
**Etching, dislocation,** definition ....48
**Etching, dissolution** ....42
ceramic materials ....160
definition ....48
**Etching, double,** definition ....48
**Etching, drop,** definition ....48
**Etching, dry,** definition ....48
**Etching, electrochemical** ....40–43
ceramic materials ....159(F), 160–162
definition ....48
**Etching, electrolytic** ....32, 42
aluminum and aluminum alloys ....63
beryllium and beryllium alloys ....70
boride ceramics ....169
cadmium and cadmium alloys ....75
ceramic materials ....161, 167, 171–172, 174
cobalt and cobalt alloys ....79–80
copper and copper alloys ....94
germanium and germanium alloys ....108
gold and gold alloys ....67
indium and indium alloys ....75
iridium and iridium alloys ....132
iron oxide layers ....102
iron, steel, and cast iron ....100–101
lead and lead alloys ....130
magnesium and magnesium alloys ....117
nickel and nickel alloys ....125
nitride ceramics ....168
osmium ....133
palladium alloys ....132
platinum and platinum alloys ....132
plutonium and plutonium alloys ....135
refractory metals and their alloys ....87
rhenium and rhenium alloys ....132, 133
ruthenium and ruthenium alloys ....132, 133
selenium and selenium alloys ....108
semiconductors ....108
silver and silver alloys ....57
tellurium and tellurium alloys ....108
thorium ....135
tin and tin alloys ....139
titanium and titanium alloys ....143
uranium and uranium alloys ....135
zinc and zinc alloys ....151
zirconium and zirconium alloys ....112
**Etching, eutectic cell,** definition ....48
**Etching, grain-boundary,** definition ....48
**Etching, grain-contrast,** definition ....48
**Etching, hot,** definition ....48
**Etching, identification (selective)**
definition ....48
**Etching, immersion,** definition ....48
**Etching, immersion (cyclic)**
definition ....48

**Etching, intermediate,** refractory metals and their alloys ..............80, 81
**Etching, ion** ................................45–47(F)
ceramic materials ..............................162
definition ..............................................48
**Etching, long-term,** definition ..............48
**Etching, metallographic.** *See also* specific types of etching. ....38–50(F)
darkfield illumination ..........................39
deposition of interference layers ..45–47
etching reproducibility .................45–47
etching temperature ............................42
etching times .......................................42
etch rinsing, definition ........................48
field metallography ..............................50
heat tinting ..........................................42
interference contrast ...........................39
methods ...................................38–45(F)
phase contrast ......................................39
polarized light ......................................40
rate of etch attack ................................42
stability of etching solutions ...............42
staining, definition ...............................49
terminology ....................................47–50
**Etching, multiple,** definition .................49
**Etching, network,** definition .................49
**Etching, optical** ..........................39–40(F)
ceramic materials .................159–160(F)
definition ..............................................49
polymers ............................................179
**Etching, physical** ........................44–45(F)
aluminum and aluminum alloys ..........64
beryllium and beryllium alloys ...........70
ceramic materials ........................159(F), 162–163, 174
cobalt and cobalt alloys .......................80
copper and copper alloys ....................94
definition ..............................................49
iron, steel, and cast iron ....................102
mercury and amalgams .....................114
nickel and nickel alloys .....................125
rare earth metals and their alloys ....................................147–148
titanium and titanium alloys .............144
zinc and zinc alloys ...........................151
**Etching, plasma**
ceramic materials ..............................161
definition ..............................................49
**Etching, potentiostatic** ..............43, 45–47
definition ..............................................49
**Etching, precipitation** ...........................42
definition ..............................................49
**Etching, primary,** definition .................49
**Etching, print (printing),** definition .....49
**Etching, reproducible** .............. 43, 44(F)
**Etching, secondary,** definition ..............49
**Etching, segregation (coring)**
definition ..............................................49
**Etching, short-term,** definition .............49
**Etching, shrink,** definition ....................49
**Etching, strain,** definition .....................49
**Etching, thermal** ..................................45
ceramic materials ..............................162
definition ..............................................50
**Etching, wet,** definition .........................50
**Ethanol**
disposal safety and regulations .........204
formula, hazards and properties ........191
safety precautions with .....................201
**Ether,** as rinse for drying the specimen surface .............................38
**Ethylenediaminetetra acetic acid (EDTA),** formula, hazards and properties ......................................191
**Ethylene glycol**
disposal safety and regulations .........204
formula, hazards and properties ........192
**Europium and europium alloys,** etching techniques and preparation of surfaces. *See also* Rare earth metals and their alloys. ..................................144–148
**Explosive,** hazardous chemical symbol and its interpretation ........198

## F

**Failure analysis quality control** ........1(F)
**Ferrous oxide coatings**
for aluminum alloys ............................64
**Filing** ......................................................18
**Filling effect** .....................................21(F)

**Flammable, highly,** hazardous chemical symbol and its interpretation....197
**Flammable, light,** hazardous chemical symbol and its interpretation....197
**Flick etchant,** composition and application guidelines....187
**Flowed material**....8
**Fluoroboric acid,** formula, hazards and properties....192
**Forced corrosion**....41
**Formic acid**
dangerous chemical reactions....202
formula, hazards and properties....192
**Fractography**....1(F)
**Freezing technique,** polymers....182
**Fry etchants,** composition and application guidelines....187

## G

**Gadolinium and gadolinium alloys,** etching techniques and preparation of surfaces. *See also* Rare earth metals and their alloys....144–148
**Galvanization**....16
**Germanium and germanium alloys.** *See also* Semiconductors.
chemical polishing....103, 107
electrolytic etching....108
electrolytic polishing....107
etching techniques and preparation of surfaces....103–108
lapping....103
macroetching....107
microetching....107–108
polishing....103, 107
preparation of surfaces....103
**Ghost structures**
cadmium and cadmium alloys....73
indium and indium alloys....73
lead and lead alloys....126
thallium and thallium alloys....73
tin and tin alloys....135
zinc and zinc alloys....148
**Glacial acetic acid**
dangerous chemical reactions....202
**Glacial acetic acid ≥96%,** formula, hazards and properties....192
**Glycerine,** formula, hazards and properties....192
**Glycerol,** formula, hazards and properties....192
**Glycol,** formula, hazards and properties....192
**Gold and gold alloys**
electrochemical polishing....64, 66
electrolytic etching....67
electrolytic lapping....64, 65, 66
electrolytic polishing....64, 65, 66
etching techniques and preparation of samples....64–67
grinding....64, 65
lapping....65
macroetching....66
microetching....66–67
polishing....64, 65, 66
preparation of samples....64–66
ultramilling....64, 65
**Gold dental alloys,** etching techniques and preparation of samples. *See also* Gold and gold alloys....65
**Gold precious metal coatings,** etching techniques and preparation of samples. *See also* Gold and gold alloys....65–66
**Gold solders,** etching techniques and preparation of samples. *See also* Gold and gold alloys....65
**Gold (III) chloride**
formulas, hazards and properties....192
**Grain boundaries**....2
**Grinding**....8, 18–37(F,T)
abrasive particle size....19–20
aluminum and aluminum alloys....57, 59
aluminum metal-matrix composites....59
americium and americium alloys....133
automatic....28
beryllium and beryllium alloys....67, 68, 69
bismuth and bismuth alloys....70, 71

**Grinding (continued)**
cadmium and cadmium alloys ............73
ceramic materials.............154, 155–158, 163(T), 164(T), 173
coarse ...............................18–19, 20, 21
cobalt and cobalt alloys...........75, 76, 77
compared to polishing.........................20
compounds and their application...22(T)
copper and copper alloys ...................89
dry ...............................................25–26
electrolytic ..............................30–33(F)
electrolytic polishing ..........................50
fine ..............................................18–19
gold and gold alloys .....................64, 65
hafnium and hafnium alloys......109, 110
indium and indium alloys ..................73
iron, steel, and cast iron...........95, 96, 97
lanthanum and lanthanum alloys....................................144–145
lead and lead alloys...........125, 126, 127
lubricants for................................25–26
magnesium and magnesium alloys ....................................114, 115
manganese and manganese alloys.....118
mechanical ...........................19–29(F,T)
mercury and amalgams .....................113
methods.........................................19(F)
neptunium and neptunium alloys......133
nickel and nickel alloys.............119, 120
platinum metals and their alloys....................................130, 131
plutonium and plutonium alloys.......133
polymers ..................................178–179
pressure, time, and rotating speed.......27
rare earth metals and their alloys....................................144–145
refractory metals and their alloys..................................80, 81, 82
semiconductor $A_{III}B_{VI}$ compounds..................................105
semiconductors ................104, 105, 106
silicon and silicon alloys...................104
silver and silver alloys .................53, 54
specimen motion during process ........27
standard recipe for mechanical.....27–28
stresses effect on microhardness.........36
substrates ................................24–25(T)
thallium and thallium alloys ..............73
thorium and thorium alloys...............133
tin and tin alloys........................136, 137
titanium and titanium alloys .....139, 140
total depth of disturbed surface layer...............................20(F)
uranium and uranium alloys .............133
wet................................................25–26
yttrium and yttrium alloys ........144–145
zinc and zinc alloys..................148, 149
zirconium and zirconium alloys ...................................109, 110
**Grinding, automatic**
ceramic materials......................156–157
**Grinding, belt** ..................................19(F)
**Grinding, electroerosive** 19(F), 30–33(F)
**Grinding, electrolytic** ................30–33(F)
ceramic materials.............................158
**Grinding, electrolytical mechanical**................................19(F)
**Grinding, manual or hand** .............19(F)
lead and lead alloys..........................126
**Grinding, mechanical**..........................18
**Grinding, rotation**............................19(F)
**Grinding, vibration**..........................19(F)
**Grinding papers**........................22–23(T)
**Grit numbers**............................22–23(T)
**Grit size** .................................................22
**Groesbeck etchant,** composition and application guidelines..................187

## H

**Hafnium and hafnium alloys**
anodizing..........................................112
chemical-mechanical polishing...............................109, 110
chemical polishing ....................109, 110
electrochemical polishing.................110
electrolytic polishing ................109, 110
etching techniques and preparation of surfaces..........109–112
etch polishing ...................................110
grinding ...................................109, 110
macroetching....................................111
microetching.....................................111
polishing..................................109, 110

preparation of surfaces.............109–110
ultramilling .......................................109
vibratory polishing...................109, 110
**Halogen hydrocarbons**
dangerous chemical reactions...........202
**Hardmetals**
electrolytic etching ..........................174
microetching.............................173–174
physical etching ...............................174
preparation of surfaces..............172–173
**Hasson etchant,** composition and
application guidelines..... .............187
**Hazardous chemical symbols and**
**their interpretation**.............196–198
**Hazardous materials**
disposal of, legal rules and
regulations ............................203–204
rules, handling tips and
precautions............................198–199
**Health hazards,** of etchants...................53
**Heat tinting,** definition ..........................48
**Hexamethylenetetramine**
formula, hazards and properties........192
**Heyn etchant,** composition and
application guidelines...................187
**High-temperature-resistant**
**cobalt-base alloys (superalloys)**
etching techniques and
preparation of surfaces..76, 78, 79, 80
**High-temperature-resistant**
**nickel-base alloys (superalloys)**
etching techniques and
preparation of surfaces..........120–125
**Holmium and holmium**
**alloys,** etching techniques and
preparation of surfaces. *See also*
Rare earth metals and their
alloys. ..................................144–148
**Hydrocarbons,** disposal safety
and regulations.............................204
**Hydrochloric acid**
dangerous chemical reactions...........203
formula, hazards and properties........192
**Hydrocyanic acid (HCN)**
safety precautions with ....................201
**Hydrofluoric acid**
dangerous chemical reactions...........203
formula, hazards and properties........192
safety precautions with ....................201
**Hydrogen**
formula, hazards and properties........192
**Hydrogen electrode**..............................41
**Hydrogen peroxide**
as chemical polishing solution............34
dangerous chemical reactions...........203
for electrolytic lapping........................36
formula, hazards and properties........192
safety precautions with ....................201

## I

**Ideal single crystal** .................................2
**Identification (marking)**.............8, 18(F)
aluminum and aluminum alloys..........58
ceramic materials .............................155
polymers..........................................177
**Indium and indium alloys**
chemical polishing........................73, 74
electrochemical polishing...................74
electrolytic etching .............................75
electrolytic polishing ....................73, 74
etching techniques and
preparation of samples.............72–75
grinding..............................................73
machining...........................................73
microetching ................................74–75
polishing.......................................73, 74
preparation of samples.................72–74
ultramilling .........................................73
**Infiltration,** ceramic materials............155
**Inflammable substances,** safety
precautions with. *See also*
Flammable. .................................202
**Inscribing**........................................18(F)
**Interference layer contrast**..................44
**Interference layers** ...............................44
ceramic materials .............................162
**Iodine**
dangerous chemical reactions...........202
formula, hazards and properties........192
**Iridium and iridium alloys.** *See*
*also* Platinum metals and
their alloys.
electrolytic etching ...........................132

**Iridium and iridium alloys (continued)**
etching techniques and
preparation of surfaces..........130–133
microetching ....................................132
preparation of surfaces.............130–131
**Iron.** *See also* Iron, steel, and cast iron.
etching techniques and
preparation of surfaces...........94–102
iron oxide layers on..................101–102
preparation of surfaces.................94–96
**Iron, steel, and cast iron**
electrochemical polishing...................96
electrolytic etching ..................100–101
electrolytic polishing ....................95, 96
grinding .................................95, 96, 97
lapping...............................................96
macroetching.......................................97
microetching ..............................98–100
physical etching ...............................102
polishing......................................95, 96
rust layers' removal ..........................102
**Iron oxide**..........................................22(T)
**Iron oxide layers**
electrolytic etching ...........................102
on iron ..............................................101
**Iron (II) sulfate**
formula, hazards and properties........192
**Iron (III) chloride**
formula, hazards and properties........192
**Iron (III) nitrate**
formula, hazards and properties........192
**Irritating,** hazardous chemical
symbol and its interpretation ........197

## K

**Kalling 1 etchant,** composition and
application guidelines..................187
**Kalling 2 etchant,** composition and
application guidelines..................187
**Keller etchants,** composition and
application guidelines..................188
**Kerr effect**
cobalt magnetic alloys ........................80
magneto-optical ..........................40, 147
**Kieselguhr and Tripoli** ....................22(T)
**Kingswater**
formula, hazards and properties.......190
**Klemm etchant,** composition and
application guidelines..................188
**Klemm I etchant,** composition and
application guidelines..................188
**Klemm II etchant,** composition and
application guidelines..................188
**Klemm III etchant,** composition and
application guidelines..................188
**Köhler's illumination principle** ...........39
**Kroll etchant,** composition and
application guidelines..................188

## L

**Lacquer**
transparent, to preserve surfaces.........43
**Lactic acid**
formula, hazards and properties........192
**Lanthanum and lanthanum alloys**
anodizing...........................................145
chemical polishing............................145
electrochemical polishing.................145
etching techniques and
preparation of surfaces..........144–148
flammability......................................144
grinding...................................144–145
microetching ....................................147
polishing...........................................145
preparation of surfaces.............144–146
sectioning.........................................144
**Lapping**....................................18, 28–29
aluminum and aluminum alloys..........60
aluminum metal-matrix composites ...60
ceramic materials....................157, 158,
163(T), 164(T), 173
cobalt and cobalt alloys.......................77
electrolytic ..............................35–36(F)
germanium and germanium alloys ...103
gold and gold alloys ............................65
iron, steel, and cast iron ......................96
lubricants for .......................................29
selenium and selenium alloys ...........104
semiconductor $A_{II}B_{VI}$ compounds....105

semiconductor $A_{III}B_{VI}$
compounds ........104–105
semiconductors ........103, 104, 105, 106
silicon and silicon alloys ........104
tellurium, pure ........104
**Lapping, electrolytic**
aluminum and aluminum alloys ........61
beryllium and beryllium alloys ........69
copper and copper alloys ........90, 91
gold and gold alloys ........64, 65, 66
nickel and nickel alloys ........121
platinum metals and their alloys ........132
refractory metals and their alloys ..81, 83
silver and silver alloys ........54, 55
**Lapping disks** ........29
**Lattice defects** ........1–2
according to geometric extension ....2(T)
nonequilibrium ........2
structural ........2
**Layer composite materials**
preparation of surfaces ........154
**Lead acetate**
formula, hazards and properties ........192
**Lead and lead alloys**
electrochemical polishing ........127
electrolytic etching ........130
electrolytic polishing ........126, 127
etching techniques and
preparation of surfaces ........125–130
grinding ........125, 126, 127
lubricants ........126, 127
macroetching ........128
microetching ........128–129
micromilling ........126
mounting ........126
polishing ........126, 127
preparation of surfaces ........125–127
rotational polishing ........126
sectioning ........126
toxicity ........125
ultramilling ........126
vibratory polishing ........126
**Lichtenegger and Bloech etchant**
composition and application
guidelines ........188
**Light microscopy,** magnification and
resolution ranges ........4(F)
**Light optical microscopy** ........3, 4
**Light poison**
hazardous chemical symbol and
its interpretation ........196
**Lipowitz's alloy**
composition and melting point ........17
**Literature references** ........205–212
ceramic materials ........211–212
compilations ........208
general textbooks ........205–208
handbooks ........208
journals ........205
metallic materials ........210–211
metallographic preparation ........209–210
polymer materials ........212
reviews ........205–208
tables ........208
**Lithium hydroxide**
formula, hazards and properties ........192
**Lubricants** ........53
for ceramic materials ........157–158
for grinding ........25–26
for lapping ........29
for lead and lead alloys ........126, 127
for polishing rare earth metals
and their alloys ........145, 146
for polymers ........177
inert ........9
**Lutetium and lutetium alloys,** etching
techniques and preparation of
surfaces. *See also* Rare earth
metals and their alloys. ........144–148

# M

**Machining**
cadmium and cadmium alloys ........73
indium and indium alloys ........73
thallium and thallium alloys ........73
**Macroetching**
aluminum and aluminum alloys ........61
antimony and antimony alloys ......71–72
beryllium and beryllium alloys ........69
bismuth and bismuth alloys ........71–72
cobalt and cobalt alloys ........78
copper and copper alloys ........92
definition ........48

**Macroetching (continued)**
germanium and germanium alloys....107
gold and gold alloys ............................66
hafnium and hafnium alloys..............111
iron, steel, and cast iron ......................97
lead and lead alloys...........................128
magnesium and magnesium
alloys .............................................116
nickel and nickel alloys.....................122
platinum and platinum alloys............132
refractory metals and their alloys .......85
semiconductors .................................107
silver and silver alloys ................56, 107
tin and tin alloys ................................137
titanium and titanium alloys .....141–142
uranium and uranium alloys .............134
zinc and zinc alloys...........................150
zirconium and zirconium alloys........111
**Macroscopy**........................................1(F)
**Magnesia**..........................................22(T)
as abrasive..........................................24
**Magnesium and magnesium alloys**
chemical-mechanical polishing ........115
chemical polishing ............114, 115, 116
electrochemical polishing.........115–116
electrolytic etching............................117
electrolytic polishing.................114, 115
etching techniques and
preparation of surfaces..........114–117
flammability.......................................114
grinding ...................................114, 115
macroetching.....................................116
microetching ............................116–117
mounting ..........................................114
polishing...................................114, 115
preparation of surfaces.............114–115
ultramilling........................................114
**Magnesium oxide**
formula, hazards and properties........192
**Magnesium perchlorate**
formula, hazards and properties........192
**Magnification,** ranges of common
optical equipment.........................4(F)
**Manganese and manganese alloys**
electrochemical polishing.................118
electrolytic polishing ........................118
etching techniques and
preparation of surfaces..........117–118
grinding ............................................118
microetching .....................................118
polishing...........................................118
preparation of surfaces.............117–118
**Marble etchants,** composition and
application guidelines...................188
**Materialography** ....................................3
**Material tolerances** ..............................3
**Melting,** partial........................................9
**Melting point,** as microstructure-
insensitive property.......................2–3
**Mercury and amalgams**
etching techniques and
preparation of surfaces..........112–114
etch polishing ...................................113
grinding ............................................113
microetching ............................113–114
mounting ..........................................112
physical etching ...............................114
polishing...........................................113
preparation of surfaces.............112–113
reactive sputtering............................114
toxicity ..............................................112
ultramilling........................................113
**Mercury (II) nitrate**
formula, hazards and properties........192
**Mesh number**.........................................22
**Metal ceramics.** *See also* Ceramic
materials.
polishing...........................................158
preparation of surfaces ................164(T)
**Metallographic examination**..................3
**Metallography** .......................................3
cleaning..................................37–38(F)
field metallography..................50–52(T)
identification (marking).................18(F)
mounting ..............................11–17(F,T)
nondestructive metallography 50–52(F)
preparation and etching recipes
for metallic materials..............52–151
preparation of specimens ..................7–8
sectioning of specimens .............8–11(F)
**Methanol**
disposal safety and regulations.........204
formula, hazards and properties........192
safety precautions with .....................201

**Methyl methacrylate**
formula, hazards and properties........192
**Microetching**
aluminum and aluminum alloys ...62–63
americium alloys.............................135
antimony and antimony alloys............72
beryllium and beryllium alloys.....69–70
bismuth and bismuth alloys ................72
boride ceramics ................................169
cadmium and cadmium alloys ......74–75
carbide ceramics ......................169–171
ceramic materials.....................164–167, 172, 173–174
cobalt and cobalt alloys.................78–79
copper and copper alloys .............92–94
definition ............................................49
germanium and germanium alloys...................................107–108
gold and gold alloys ......................66–67
hafnium and hafnium alloys.............111
indium and indium alloys .............74–75
iridium..............................................132
iron, steel, and cast iron .............98–100
lanthanum and lanthanum alloys ......147
lead and lead alloys..................128–129
magnesium and magnesium alloys ...................................116–117
manganese and manganese alloys.....118
mercury and amalgams .............113–114
nickel and nickel alloys ............122–124
nitride ceramics................................168
osmium............................................132
palladium and palladium alloys........132
platinum and platinum alloys............132
plutonium alloys ..............................135
rare earth metals and their alloys ......147
refractory metals and their alloys 85–87
rhenium ............................................132
ruthenium .........................................132
selenium and selenium alloys ...........108
semiconductor $A_{II}B_{VI}$ and $A_{III}B_{VI}$ compounds ......................................108
semiconductors .........................107–108
silicon and silicon alloys...........107–108
silver and silver alloys .................56–57
tellurium and tellurium alloys...........108
thallium and thallium alloys .........74–75
thorium and thorium alloys..............134
tin and tin alloys ..............................138
titanium and titanium alloys .....142–143
uranium and uranium alloys .....134–135
zinc and zinc alloys ..................150–151
zirconium and zirconium alloys........111
**Microhardness,** grinding and polishing stresses' effect.................36
**Micromilling,** lead and lead alloys ......126
**Microprobe analysis**..............................11
**Microstructural design**...........................3
**Microstructural engineering**.................3
**Microstructure** ...................................1–2
definition ..............................................2
**Microstructure-insensitive properties** ...................................2–3
**Microstructure microscopy**..............1(F)
**Microstructure-sensitive properties** ...................................2–3
**Microtome cutting** ..........................29–30
ceramic materials .............................158
**Microtome sectioning**
polymers ............................182–183(T)
**Microtome ultramilling**.................29–30
**Microtoming** ..........................................18
diamond-knife...................................36
electrolytic ..........................................36
tin and tin alloys ...............................136
titanium and titanium alloys .............139
zinc and zinc alloys ...........................148
**Milling,** fine...........................................18
**Mint gold,** etching techniques and preparation of samples..............64–65
**Mohr's salt,** formula, hazards and properties ....................................190
**Molten salts,** safety precautions with ..202
**Molybdenum and molybdenum alloys,** etching techniques and preparation of surfaces. *See also* Refractory metals and their alloys. ..............................80–88
**Molybdic acid**
formula, hazards and properties........193
**Mounting**.............................8, 11–17(F,T)
antimony and antimony alloys............70
bismuth and bismuth alloys ................70
cadmium and cadmium alloys ............73
ceramic materials......................154–155
clamping..................................11–13(F)

**Mounting (continued)**
cold mounting ..............13(F), 14(T), 15
comparison of materials ................14(T)
considerations ....................................11
elastic spacers for clamping................13
embedding...............11, 12(F), 13–15(F)
for electrolytic sample
preparation ...............................16–17
gold and gold alloys ............................64
hot mounting .................13(F), 14(T), 15
indium and indium alloys ...................73
infiltration ..........................................17
lead and lead alloys ...........................126
magnesium and magnesium alloys ...114
mercury and amalgams .....................112
methods of...............................11–12(F)
mounting media..................................11
polymers...........................................177
of powders..........................................17
properties of materials ...................14(T)
requirements of materials .............13–15
semiconductors .........................103, 106
techniques ...............................15–17(F)
thallium and thallium alloys ...............73
tin and tin alloys ................................135
titanium and titanium alloys .............139
**Murakami etchant,** composition
and application guidelines ............188

## N

**Neodymium and neodymium alloys,** etching techniques and preparation of surfaces. *See also* Rare earth metals and their alloys. ..........................144–148
**Neptunium and neptunium alloys**
electrochemical polishing.................133
electrolytic polishing ........................133
etching techniques and
preparation of surfaces..........133–135
grinding ............................................133
polishing...........................................133
preparation of surfaces......................133
radioactivity .....................................133
**Newton's alloy,** composition and
melting point ..................................17
**Nickel and nickel alloys**
chemical polishing............................121
electrochemical polishing.................121
electrolytic etching ...........................125
electrolytic lapping ...........................121
electrolytic polishing ........119, 120, 121
etching techniques and
preparation of surfaces..........118–125
grinding ............................119, 120, 122
macroetching....................................122
microetching ............................122–124
physical etching ...............................125
polishing...................................119, 120
preparation of surfaces.............118–120
shock polishing .........................119, 120
toxicity ..............................................118
vibratory polishing............................119
**Niobium and niobium alloys,** etching techniques and preparation of surfaces. *See also* Refractory metals and their alloys. ...............................80–88
**Nital etchant,** composition and
application guidelines ...................188
**Nitric acid**
as chemical polishing solution............34
dangerous chemical reactions...........203
for electrolytic lapping........................36
formula, hazards and properties........193
safety precautions with .....................201
**Nitride ceramics.** *See also* Ceramic materials.
electrolytic etching ...........................168
microetching ....................................168
**Nitrogen**
formula, hazards and properties........193
**Nitrohydrochloric acid**
formula, hazards and properties........190

## O

**Oberhoffer etchant,** composition
and application guidelines ............189
**Oberhoffer method** ...............................43

**Oil-immersion objective lenses**..........160
**Orange-peel effect**..................................34
**Orthophosphoric acid**
formula, hazards and properties........193
**Osmium and osmium alloys.** *See also* Platinum metals and their alloys.
electrolytic etching ...........................133
etching techniques and preparation of surfaces..........130–133
microetching .....................................132
preparation of surfaces..............130–131
sectioning.........................................131
**Oxalic acid**
dangerous chemical reactions...........203
for electrolytic lapping.........................36
formula, hazards and properties........193
**Oxide ceramics.** *See also* Ceramic materials.
electrolytic etching ...........................167
microetching ............................164–167
**Oxide-dispersion-strengthened (ODS) superalloys,** etching techniques and preparation of surfaces ..............................120–125
**Oxidizing materials,** hazardous chemical symbol and its interpretation.................................197

# P

**Palladium and palladium alloys.** *See also* Platinum metals and their alloys.
electrolytic etching ...........................132
etching techniques and preparation of surfaces..........130–133
microetching .....................................132
preparation of surfaces..............130–131
**Palmerton etchants,** composition and application guidelines ............189
**Particle size**................................22–23(T)
**Perchloric acids** ..............................30–31
dangerous chemical reactions...........203
formula, hazards and properties........193
safety precautions for work 199–201(F)
ternary diagram ...........................200(F)
**Peroxide**
dangerous chemical reactions...........203
**Phase contrast,** polymers ....................183
**Phenol resins**
as mounting materials....................14(T)
properties ......................................14(T)
**Phosphide ceramics,** microetching. *See also* Ceramic materials. ................172
**Phosphoric acids** ..................................30
as viscous agent for chemical polishing.........................................34
**Photographic fixer**
formula, hazards and properties........195
**Picral etchant,** composition and application guidelines ...................189
**Picrates**
dangerous chemical reactions...........203
**Picric acid**
dangerous chemical reactions...........203
for electrolytic lapping.........................36
formula, hazards and properties........193
**Pioloform F50 replica technique,** material used, application technique, and microscopic resolution ..............51(T)
**Planing** ..................................................18
**Plastographic examination** ....................3
**Plastography** .........................175–184(T)
definition ...........................................175
**Platinum and platinum alloys.** *See also* Platinum metals and their alloys.
electrolytic etching ...........................132
etching techniques and preparation of surfaces..........130–133
macroetching....................................132
microetching ....................................132
preparation of surfaces..............130–131
**Platinum metals (platinum, palladium, rhodium, iridium, ruthenium and osmium) and their alloys**
electrochemical polishing.................132
electrolytic lapping ...........................132
etching techniques and preparation of surfaces..........130–133
etch polishing...................................131
grinding...................................130, 131
polishing..................................130, 131

**Platinum metals and their alloys (continued)**
sectioning....130
vibratory polishing....131
**Plucking**....8
**Plutonium alloys,** microetching..134–135
**Plutonium and plutonium alloys**
electrochemical polishing....133
electrolytic etching....135
electrolytic polishing....133
etching techniques and preparation of surfaces....133–135
grinding....133
polishing....133
preparation of surfaces....133
radioactivity....133
**Polarization,** ceramic materials....160
**Polarized light,** polymers....183
**Polishing**....8, 18–37(F,T)
abrasive particle size....19–20
aluminum and aluminum alloys....58, 59, 60, 64
aluminum metal-matrix composites...60
americium and americium alloys....133
automatic....28
beryllium and beryllium alloys....67, 68–69
bismuth and bismuth alloys....71
cadmium and cadmium alloys....73, 74
ceramic materials....155–158(F), 163(T), 164(T), 173
cloths available....25, 26(T)
coarse....19
cobalt and cobalt alloys....75, 76, 77
combined methods of polishing....34–36(F)
compared to grinding....20
compounds and their application...22(T)
copper and copper alloys....89–90
electrolytic....30–33(F)
evaluation of methods....36–37(F)
germanium and germanium alloys....103, 107
gold and gold alloys....64, 65, 66
hafnium and hafnium alloys....109, 110
indium and indium alloys....73, 74
iron, steel, and cast iron....95, 96
lanthanum and lanthanum alloys....145, 146
lead and lead alloys....126, 127
magnesium and magnesium alloys....114, 115
manganese and manganese alloys....118
mechanical....19–29(F,T)
mercury and amalgams....113
methods....19(F)
neptunium and neptunium alloys....133
nickel and nickel alloys....119, 120
platinum metals and their alloys....130, 131
plutonium and plutonium alloys....133
polymers....178–179(T)
pressure, time, and rotating speed....27
rare earth metals and their alloys....145, 146
refractory metals and their alloys....80, 81–82
selenium and selenium alloys....104
semiconductor $A_{II}B_{VI}$ compounds....105
semiconductor $A_{III}B_{VI}$ compounds....104–105
semiconductors....103, 104, 105, 106, 107
silicon and silicon alloys....104
silver and silver alloys....54–55
specimen motion during process....27
standard recipe for mechanical....27–28
stresses effect on microhardness....36
substrates....24–25(T)
tellurium, pure....104
thallium and thallium alloys....73, 74
thorium and thorium alloys....133
tin and tin alloys....136, 137
titanium and titanium alloys....139–140
total depth of disturbed surface layer....20(F)
uranium and uranium alloys....133
yttrium and yttrium alloys....145, 146
zinc and zinc alloys....148, 149
zirconium and zirconium alloys 109, 110
**Polishing, anodic**....30–33(F)
multiphase alloys....33
**Polishing, attack**....35
definition....47
refractory metals and their alloys....80

**Polishing, automatic**
aluminum and aluminum alloys..........59
aluminum metal-matrix composites...59
silver and silver alloys ......................54
**Polishing, chemical**...............18, 19(F), 34
advantages..........................................34
aluminum and aluminum alloys....58, 60
cadmium and cadmium alloys ......73, 74
ceramic materials.............................158
cobalt and cobalt alloys.................77–78
copper and copper alloys .............90, 91
disadvantages......................................34
germanium and germanium
alloys ....................................103, 107
hafnium and hafnium alloys......109, 110
indium and indium alloys.............73, 74
lanthanum and lanthanum alloys......145
magnesium and magnesium
alloys ............................114, 115, 116
nickel and nickel alloys....................121
rare earth metals and their alloys ......145
refractory metals and their alloys 83–84
semiconductors ................103, 104, 107
silicon and silicon alloys...................104
silver and silver alloys ..................54, 55
thallium and thallium alloys .........73, 74
tin and tin alloys...............................137
titanium and titanium alloys .....140, 141
yttrium and yttrium alloys ................145
zinc and zinc alloys..................149–150
zirconium and zirconium
alloys ....................................109, 110
**Polishing, chemical-mechanical** ..........35
hafnium and hafnium alloys......109, 110
magnesium and magnesium alloys...115
rare earth metals and their alloys ......146
refractory metals and their
alloys ............................80, 81, 84–85
zirconium and zirconium
alloys ....................................109, 110
**Polishing, electrochemical**
aluminum and aluminum alloys ...60–61
americium and americium alloys......133
beryllium and beryllium alloys.....68–69
cadmium and cadmium alloys ............74
cobalt and cobalt alloys.................77–78
copper and copper alloys .............90–91
gold and gold alloys .....................64, 66
hafnium and hafnium alloys.............110
indium and indium alloys ..................74
iron, steel, and cast iron ......................96
lanthanum and lanthanum alloys ......145
lead and lead alloys ..........................127
magnesium and magnesium
alloys ....................................115–116
manganese and manganese alloys.....118
neptunium and neptunium alloys......133
nickel and nickel alloys.....................121
platinum metals and their alloys .......132
plutonium and plutonium alloys .......133
rare earth metals and their
alloys ....................................145, 146
refractory metals and their alloys 82–85
semiconductors ................................107
silver and silver alloys .......................55
thallium and thallium alloys ..............74
thorium and thorium alloys...............133
tin and tin alloys.......................136–137
titanium and titanium alloys ............141
uranium and uranium alloys ............133
zinc and zinc alloys..................149–150
zirconium and zirconium alloys........110
**Polishing, electrolytic**.....................11, 18,
19(F), 30–33(F), 50
advantages..........................................33
aluminum and aluminum
alloys ............................58, 59, 60, 64
americium and americium alloys......133
beryllium and beryllium alloys.....67, 68
bismuth and bismuth alloys ................71
cadmium and cadmium alloys ......73, 74
ceramic materials.............................158
cobalt and cobalt alloys...........75, 76, 77
copper and copper alloys ..............90–91
germanium and germanium alloys ...107
gold and gold alloys ...............64, 65, 66
hafnium and hafnium alloys......109, 110
indium and indium alloys..............73, 74
iron, steel, and cast iron ...............95, 96
lead and lead alloys ..................126, 127
limitations ..........................................33
magnesium and magnesium
alloys ....................................114, 115
manganese and manganese alloys.....118
neptunium and neptunium alloys......133
nickel and nickel alloys.....119, 120, 121

**Polishing, electrolytic (continued)**
plutonium and plutonium alloys.......133
prerequisite .......................................33
refractory metals and their
alloys.................................81, 82–83
semiconductors .......................104, 107
silicon and silicon alloys..................104
silver and silver alloys .................54, 55
thallium and thallium alloys .........73, 74
thorium and thorium alloys...............133
tin and tin alloys.......................136–137
titanium and titanium alloys .....139, 141
uranium and uranium alloys .............133
zinc and zinc alloys..........................149
zirconium and zirconium alloys 109, 110
**Polishing, electrolytical mechanical**...............................19(F)
**Polishing, etch**.......................................35
ceramic materials......................161–162
copper and copper alloys ....................89
hafnium and hafnium alloys.............110
mercury and amalgams .....................113
platinum metals and their alloys .......131
rare earth metals and their
alloys...................................145, 146
zinc and zinc alloys...................148–149
zirconium and zirconium alloys........110
**Polishing, manual**............................19(F)
final, silver and silver alloys ..............54
**Polishing, mechanical**...........................18
**Polishing, mechanical-chemical** .....19(F)
**Polishing, multiple**.........................34–35
**Polishing, rotational**........................19(F)
lead and lead alloys ..........................126
**Polishing, shock**....................................33
aluminum and aluminum alloys....58, 59
beryllium and beryllium alloys...........68
cobalt and cobalt alloys.......................76
nickel and nickel alloys............119, 120
refractory metals and their alloys .......81
titanium and titanium alloys .....139, 140
**Polishing, vibratory** ..................19(F), 28
beryllium and beryllium alloys...........68
bismuth and bismuth alloys ................71
hafnium and hafnium alloys......109, 110
lead and lead alloys ..........................126
nickel and nickel alloys.....................119
platinum metals and their alloys .......131
refractory metals and their alloys .......81
titanium and titanium alloys .............139
zirconium and zirconium alloys 109, 110
**Polishing cloths** .........................25, 26(T)
**Polyester resins**
as mounting materials...................14(T)
properties ......................................14(T)
**Polymers**
brightfield illumination.....................183
cast liquid film technique..................183
chemical etching .......................179–180
cleaning..............................179, 180(T)
darkfield illumination .......................179
definition ..........................................175
differential interference contrast ......179
etchants for..............................180–181
freezing technique............................182
grinding...................................178–179
identification ....................................177
liquids causing partial solution
or swelling ...............................180(T)
lubricants for ....................................177
microtome sectioning ..........182–183(T)
mounting ..........................................177
optical etching..................................179
phase contrast...................................183
polarized light ..................................183
polishing ............................178–179(T)
preparation and evaluation
criteria for light microscopic
observations ............................176(T)
preparation of surfaces ........177–179(T)
production of....................................175
properties .........................................175
reflectivity.........................................179
sampling...........................................177
softening and maximum
high-temperature use ...............177(T)
thin sectioning..........................181–182
types, groups, and their
abbreviations ..........................176(T)
ultramilling ..............................179, 182
**Porosity,** ceramic materials..................158
**Porous and brittle ceramics,** preparation of surfaces.
*See also* Ceramic materials. ....163(T)
**Potassium carbonate**
formula, hazards and properties........193

**Potassium chlorate**
formula, hazards and properties........193
**Potassium chloride**
formula, hazards and properties........193
**Potassium chromate**
formula, hazards and properties........193
**Potassium cyanide**
formula, hazards and properties........193
safety precautions with .....................201
**Potassium dichromate**
formula, hazards and properties........193
**Potassium disulfate**
formula, hazards and properties........193
**Potassium ferricyanide**
formula, hazards and properties........193
**Potassium ferrocyanide**
formula, hazards and properties........193
**Potassium hydrogen carbonate**
formula, hazards and properties........193
**Potassium hydrogen fluoride**
formula, hazards and properties........193
**Potassium hydrogen sulfate**
formula, hazards and properties........193
**Potassium hydroxide**
dangerous chemical reactions...........203
formula, hazards and properties........194
safety precautions with .....................201
**Potassium hydroxide solution**
formula, hazards and properties........194
**Potassium iodide**
formula, hazards and properties........194
**Potassium nitrate**
formula, hazards and properties........194
**Potassium permanganate**
dangerous chemical reactions...........203
formula, hazards and properties........194
**Potassium phthalate**
formula, hazards and properties........194
**Potassium thiocyanate**
formula, hazards and properties........194
**Potentiometer** ........................................43
**Powders**
inorganic, as abrasives ........................22
mounting of........................................17
**Praseodymium and prasiodymium alloys**
etching techniques and preparation of surfaces. *See also* Rare earth metals and their alloys. .....................144–148

**Preparation of surfaces**
aluminum and aluminum alloys ...57–60
americium and americium alloys......133
antimony and antimony alloys......70–71
beryllium and beryllium alloys.....67–68
bismuth and bismuth alloys ..........70–71
cadmium and cadmium alloys ......72–74
cast iron ........................................94–96
ceramic materials................154, 163(T), 164(T), 172–173
cobalt and cobalt alloys.................75–77
copper and copper alloys ..............88–90
germanium and germanium alloys ...103
gold and gold alloys ......................64–66
hafnium and hafnium alloys .....109–110
indium and indium alloys .............72–74
iridium and iridium alloys ........130–131
iron ................................................94–96
lanthanum and lanthanum alloys....................................144–146
lead and lead alloys...................125–127
magnesium and magnesium alloys ....................................114–115
manganese and manganese alloys ....................................117–118
mercury and amalgams .............112–113
neptunium and neptunium alloys......133
nickel and nickel alloys.............118–120
osmium and osmium alloys ......130–131
palladium and palladium alloys...................................130–131
platinum and platinum alloys ...130–131
plutonium and plutonium alloys .......133
printed circuits and printed circuit boards...........................................106
rare earth metals and their alloys 144–146
refractory metals and their alloys 80–82
resistors ....................................106–107
rhodium and rhodium alloys.....130–131
ruthenium and ruthenium alloys 130–131
selenium and selenium alloys ...........104
semiconductors .........................103–107
silicon and silicon alloys...................104
silver and silver alloys ..................53–55
steel ...............................................94–96
tellurium and tellurium alloys...104, 105
thallium and thallium alloys .........72–74
thorium and thorium alloys...............133

**Preparation of surfaces (continued)**
titanium and titanium alloys .....139–140
transistors .....106–107
uranium and uranium alloys .....133
yttrium and yttrium alloys .....144–146
zinc and zinc alloys .....148–149
zirconium and zirconium alloys 109–110
**Printed circuits and printed circuit boards,** preparation of surfaces....106
**Pullouts** .....8
**Pumice** .....22(T)

## Q

**Quantitative image analysis** .....47
**Quartz** .....22(T)

## R

**Rare earth metals and their alloys**
anodizing .....145, 146
chemical-mechanical polishing .....146
chemical polishing .....145
electrochemical polishing .....145, 146
etch polishing .....145, 146
grinding .....144–145
lubricants for polishing .....145, 146
microetching .....147
microscopic examination .....145
physical etching .....147–148
polishing .....145, 146
preparation of surfaces .....144–146
sectioning .....144
ultramilling .....145
**Reactive sputtering** .....44–45, 46(F)
ceramic materials .....162–163
cobalt alloys .....80
mercury and amalgams .....114
**Real single crystal** .....2
**Recrystallization** .....9
**Redox potentials** .....42
**Redox process** .....40
**Reflective layer**
ceramic materials .....159–160
**Reflectivity**
percentage, to evaluate polishing methods .....36–37(F)
polymers .....179
**Reflectivity differences (R)** .....44
**Refractory metals and their alloys**
anodizing .....88
attack polishing .....80
chemical-mechanical polishing .....80, 81, 84–85
chemical polishing .....83–84
electrochemical polishing .....82–85
electrolytic etching .....87
electrolytic lapping .....81, 83
electrolytic polishing .....81, 82–83
etching techniques and preparation of surfaces .....80–88
grinding .....80, 81, 82
intermediate etching .....80, 81
macroetching .....85
microetching .....85–87
polishing .....80, 81–82
preparation of surfaces .....80–82
shock polishing .....81
vibratory polishing .....81
**Rehbinder effect** .....26
**Remelting** .....18
**Replica techniques** .....50–52(T)
**Replicating** .....52
**Resistors**
etching techniques and preparation of surfaces .....119–125
preparation of surfaces .....106–107
**Resolution,** ranges of common optical equipment .....4(F)
**Rhenium,** microetching .....132
**Rhenium and rhenium alloys.** *See also* Refractory metals and their alloys.
electrolytic etching .....132, 133
etching techniques and preparation of surfaces .....80–88
**Rhodium and rhodium alloys.** *See also* Platinum metals and their alloys.
etching techniques and preparation of surfaces .....130–133
preparation of surfaces .....130–131
**Rinsing, etch,** definition .....48
**Rose's alloy**
composition and melting point .....17

**Rotating arm across a fixed disk** .........28
**Rotating wheels** .........28
**Rust layers' removal**
iron, steel, and cast iron .........102
**Ruthenium,** microetching.........132
**Ruthenium and ruthenium alloys.** *See also* Platinum metals and their alloys.
electrolytic etching .........132, 133
etching techniques and preparation of surfaces.........130–133
preparation of surfaces.........130–131
sectioning.........131

## S

**Safety precautions**
americium, radioactivity.........133
beryllium, etching of.........67
cadmium, toxicity .........72
etching recipes .........199–202(F)
gallium arsenide (GaAs), toxicity.....105
lead, toxicity.........125
magnesium, flammability .........114
mercury and amalgams, toxicity.......112
neptunium, radioactivity.........133
nickel, toxicity.........118
plutonium, radioactivity .........133
selenium, toxicity.........104
tellurium, toxicity .........104
thallium, toxicity.........72
thorium, radioactivity .........133
uranium, radioactivity.........133
**Safranin**.........17
**Salmiac**
formula, hazards and properties.........190
**Samarium and samarium alloys**
etching techniques and preparation of surfaces. *See also* Rare earth metals and their alloys. .........144–148
**Sample preparation, for edge retention**.........15–16
**Sampling** .........8
ceramic materials.........154
polymers.........177
**Sand blasting** .........9(F)
**Sawing**.........9(F)
**Scanning electron microscopy** .........11
**Schramm etchants,** composition and application guidelines.........189
**Scratches**.........8
ceramic materials.........158
gold and gold alloys .........64
mercury and amalgams .........112
**Sectioning**.........8–11(F)
aluminum and aluminum alloys.........54
antimony and antimony alloys.........70
bismuth and bismuth alloys .........70
ceramic materials.........154, 156(F)
lanthanum and lanthanum alloys......144
lead and lead alloys.........126
osmium and osmium alloys .........131
platinum metals and their alloys.......130
rare earth metals and their alloys......144
ruthenium and ruthenium alloys.......131
semiconductors .........103, 106
tin and tin alloys.........135
titanium and titanium alloys .........139
yttrium and yttrium alloys .........144
**Sectioning, thin,** polymers .........181–182
**Selenic acid**
formula, hazards and properties........194
**Selenium and selenium alloys.** *See also* Semiconductors.
electrolytic etching .........108
etching techniques and preparation of surfaces.........103–108
lapping .........104
microetching .........108
polishing.........104
preparation of surfaces.........104
toxicity .........104
**Semiconductor $A_{II}B_{VI}$ compounds**
lapping .........105
polishing.........105
**Semiconductor $A_{III}B_{VI}$ compounds**
grinding.........105
lapping .........104–105
polishing .........104–105
**Semiconductor $A_{III}B_{VI}$ and $A_{II}B_{VI}$ compounds**
etching techniques and preparation of surfaces.........103–108
microetching .........108
preparation of surfaces.........104–105

**Semiconductors**
chemical polishing............103, 104, 107
electrochemical polishing.................107
electrolytic etching ............................108
electrolytic polishing ................104, 107
etching techniques and
preparation of surfaces..........103–108
grinding.............................104, 105, 106
lapping.......................103, 104, 105, 106
macroetching....................................107
microetching.............................107–108
mounting..........................................106
polishing............103, 104, 105, 106, 107
preparation of surfaces..............103–107
sectioning.........................................106
**Silicon and silicon alloys.** *See also* Semiconductors.
chemical polishing............................104
electrolytic polishing ........................104
etching techniques and
preparation of surfaces..........103–108
grinding.............................................104
lapping .............................................104
macroetching....................................107
microetching.............................107–108
polishing............................................104
preparation of surfaces......................104
**Silicon carbide (SiC)**........................22(T)
as abrasive...........................................24
**Silicon carbide wet-grinding papers**
grain size designation ..............22, 23(T)
**Silicon dioxide,** as abrasive....................24
**Silver**
800/silver, etching techniques and
sample preparation..........................54
835/silver, etching techniques and
sample preparation..........................54
fine and fine-grain, etching techniques
and sample preparation.............53–54
**Silver and silver alloys**
chemical polishing........................54, 55
electrochemical polishing...................55
electrolytic etching .............................57
electrolytic lapping .......................54, 55
electrolytic polishing ....................54, 55
etching techniques and
preparation of surfaces.............53–57
grinding........................................53, 54
macroetching.......................................56
microetching ................................56–57
polishing ......................................54–55
preparation of surfaces..................53–55
ultramilling ..........................................53
**Silver-base semifinished products**
etching techniques and sample
preparation .......................................54
**Silver brazing solders,** etching
techniques and surface
preparation ................................54–55
**Silver ceramic solders,** etching
techniques and surface
preparation ................................54–55
**Silver contact materials,** etching
techniques and surface
preparation ................................54–55
**Silver cyanide**
formula, hazards and properties........194
**Silver nitrate**
dangerous chemical reactions...........203
formula, hazards and properties........194
**Silver precious metal coatings**
etching techniques and
preparation of surfaces....................55
**Silver solders**
etching techniques and
preparation of surfaces..............54–55
microetching ......................................56
**Smeared layers** ..............21(F), 23, 29, 33
cadmium and cadmium alloys ............73
ceramic materials.............................156
cobalt and cobalt alloys.......................75
indium and indium alloys ...................73
lead and lead alloys..........................126
nickel and nickel alloys.....................119
thallium and thallium alloys ...............73
tin and tin alloys...............................135
zinc and zinc alloys...........................148
**Smears**.....................................................8
mercury and amalgams .....................112
silver and silver alloys ........................53
**Smoothing**..............................................18
**Sodium carbonate**
formula, hazards and properties........194
**Sodium chlorate**
formula, hazards and properties........194

**Sodium chloride**
formula, hazards and properties........194
**Sodium chromate**
formula, hazards and properties........194
**Sodium cyanide**
formula, hazards and properties........194
**Sodium dichromate**
formula, hazards and properties........194
**Sodium fluoride**
formula, hazards and properties........194
**Sodium hydrogen carbonate**
formula, hazards and properties........194
**Sodium hydrogen phosphate**
formula, hazards and properties........194
**Sodium hydroxide**
dangerous chemical reactions...........203
formula, hazards and properties........195
**Sodium hydroxide solution**
formula, hazards and properties........195
**Sodium hypochlorite**
formula, hazards and properties........195
**Sodium peroxide**
dangerous chemical reactions...........203
formula, hazards and properties........195
**Sodium sulfate**
formula, hazards and properties........195
**Sodium sulfide**
formula, hazards and properties........195
**Sodium tetraborate**
formula, hazards and properties........195
**Sodium thiocyanate**
formula, hazards and properties........195
**Sodium thiosulfate**
for electrolytic lapping.......................36
formula, hazards and properties........195
**Solvents**
disposal safety and regulations.........204
**Sputtering**........................................46(F)
**Stacking faults**........................................2
**Staining,** definition ...............................49
**Stamping**..........................................18(F)
**Steel.** *See also* Iron, steel, and cast iron.
etching techniques and
preparation of surfaces............94–102
preparation of surfaces.................94–96
**Stellite,** etching techniques and
preparation of surfaces............76–77,
78, 79, 80
**Sterling silver,** etching techniques
and sample preparation. *See also*
Silver and silver alloys. ..................54
**Strength,** as microstructure-sensitive
property........................................2–3
**Structural lattice defects** ........................2
**Sulfide ceramics,** microetching. *See
also* Ceramic materials. ................172
**Sulfuric acids**........................................30
as chemical polishing solution............34
dangerous chemical reactions...........203
formula, hazards and properties........195
safety precautions with.....................201
**Surface deformation** ..............................8
**Surface deformation depth** .................20
**Swabbing,** definition.............................49

# T

**Tantalum and tantalum alloys**
etching techniques and preparation
of surfaces. *See also* Refractory
metals and their alloys. ............80–88
**Taper sectioning**......................15, 16(F,T)
**Taper technique**.......................15, 16(F,T)
**Tartaric acid**
formula, hazards and properties........195
**Technovit 3040 replica
technique,** material used,
application technique, and
microscopic resolution .............51(T)
**Technovit 5071 replica
technique,** material used,
application technique, and
microscopic resolution .............51(T)
**Tellurium, pure**
lapping ............................................104
polishing...........................................104
**Tellurium and tellurium
alloys.** *See also* Semiconductors.
electrolytic etching ..........................108
etching techniques and
preparation of surfaces..........103–108
microetching ....................................108
preparation of surfaces.............104, 105
toxicity ............................................104
**Tempering,** local ....................................9

**Terbium and terbium alloys,** etching techniques and preparation of surfaces. *See also* Rare earth metals and their alloys. ....144–148
**Tetrafluoromethane**
formula, hazards and properties........195
**Thallium and thallium alloys**
chemical polishing........73, 74
electrochemical polishing........74
electrolytic polishing........73, 74
etching techniques and preparation of surfaces........72–75
grinding........73
machining........73
microetching........74–75
polishing........73, 74
preparation of surfaces........72–74
ultramilling........73
**Thermal conductivity**........7
as microstructure-sensitive property 2–3
**Thermal expansion coefficient,** as microstructure-insensitive property........2–3
**Thermodynamic equilibrium**........2
**Thermoplastics**
as mounting material........13, 14(T)
**Thermosettings**
as mounting material........13, 14(T)
**Thioglycolic acid**
formula, hazards and properties........195
**Thiourea**
formula, hazards and properties........195
**Thorium,** electrolytic etching........135
**Thorium and thorium alloys**
electrochemical polishing........133
electrolytic polishing........133
etching techniques and preparation of surfaces........133–135
grinding........133
microetching........134–135
polishing........133
preparation of surfaces........133
radioactivity........133
**Thulium and thulium alloys,** etching techniques and preparation of surfaces. *See also* Rare earth metals and their alloys. ........144–148
**Tin and tin alloys**
chemical polishing........137
deformation........135
deformation twins........135
electrochemical polishing........136–137
electrolytic etching........139
electrolytic polishing........136–137
etching techniques and preparation of surfaces........135–139
ghost structures........135
grinding........136, 137
macroetching........137
microetching........138
microtoming........136
mounting........135
polishing........136, 137
preparation of surfaces........135–136
sectioning........135
smear layers........135
ultramilling........136
**Tin dioxide**........22(T)
**Tin II chloride**
formula, hazards and properties........195
**Titanium and titanium alloys**
chemical polishing........140, 141
deformation........139
electrochemical polishing........141
electrolytic etching........143
electrolytic polishing........139, 141
etching techniques and preparation of surfaces........139–144
grinding........139, 140
macroetching........141–142
microetching........142–143
microtoming........139
mounting........139
physical etching........144
polishing........139–140
preparation of surfaces........139–140
sectioning........139
shock polishing........139, 140
vibratory polishing........139
**Titanium superalloys,** etching techniques and preparation of surfaces........140, 141–144
**Toluene**
disposal safety and regulations........204
formula, hazards and properties........195

**Toxic,** hazardous chemical
symbol and its interpretation ........196
**Toxic materials**
safety precautions with ....................202
**Transcopy replica technique,** material used, application technique, and microscopic resolution ..............51(T)
**Transformation temperature**
as microstructure-insensitive property........................................2–3
**Transistors**
preparation of surfaces..............106–107
**Triafol BN replica technique,** material used, application technique, and microscopic resolution ..............51(T)
**Trichloroethylene**
disposal safety and regulations .........204
formula, hazards and properties........195
**Triethylamine**
formula, hazards and properties........195
**Tucker etchant,** composition and application guidelines...................189
**Tungsten and tungsten alloys,** etching techniques and preparation of surfaces. *See also* Refractory metals and their alloys. ............80–88
**Turning,** fine ..........................................18

## U

**Ultimate tensile strength** ........................7
**Ultramilling**....................................29–30
aluminum and aluminum alloys..........59
cadmium and cadmium alloys ............73
ceramic materials.............................158
copper and copper alloys ....................89
gold and gold alloys ......................64, 65
hafnium and hafnium alloys .............109
indium and indium alloys ...................73
lead and lead alloys...........................126
magnesium and magnesium alloys ...114
mercury and amalgams .....................113
polymers..................................179, 182
rare earth metals and their alloys ......145
silver and silver alloys ........................53
thallium and thallium alloys ...............73
tin and tin alloys...............................136
zirconium and zirconium alloys .......109
**Ultramilling, microtome**..................19(F)
**Ultrasonic chiseling**............................9(F)
**Ultrasonic cleaning** ..............................37
**Uranium and uranium alloys**
electrochemical polishing.................133
electrolytic etching ...........................135
electrolytic polishing ........................133
etching techniques and preparation of surfaces..........133–135
grinding............................................133
macroetching....................................134
microetching ............................134–135
polishing...........................................133
preparation of surfaces......................133
radioactivity .....................................133

## V

**Vacancies**..................................................2
**Vanadium and vanadium alloys**
etching techniques and preparation of surfaces. *See also* Refractory metals and their alloys. ............80–88
**Vanadium oxide**
formula, hazards and properties........196
**Vapor deposition** ........................44, 46(F)
**Vernis Replic replica technique**
material used, application technique, and microscopic resolution........51(T)
**Vibration engraving**........................18(F)
**Vickers hardness** .................................24
**Villela and Beregekoff etchant**
composition and application guidelines.....................................189
**Villela etchant,** composition and application guidelines...................189
**Viscous agents**.......................................34

## W

**Wallner etchant,** composition and application guidelines...................189
**Weck etchant,** composition and application guidelines...................189

**White gold,** etching techniques and preparation of samples. *See also* Gold and gold alloys. ....................65
**Wipe etching.** *See* Swabbing.
**Wood's metal**
composition and melting point ...........17

## X

**Xantopren replica technique,** material used, application technique, and microscopic resolution ..............51(T)
**Xylene**
disposal safety and regulations .........204
formula, hazards and properties........196

## Y

**Yellow gold,** etching techniques and preparation of samples. *See also* Gold and gold alloys. ....................65
**Yield strength** ...........................................7
**Young's modulus,** as microstructure-insensitive property.......................2–3
**Ytterbium and ytterbium alloys**
etching techniques and preparation of surfaces. *See also* Rare earth metals and their alloys. ........144–148
**Yttrium and yttrium alloys**
chemical polishing ............................145
etching techniques and preparation of surfaces..........144–148
grinding..................................144–145
polishing..................................145, 146
preparation of surfaces.............144–146
sectioning.........................................144

## Z

**Zinc and zinc alloys**
chemical polishing....................149–150
deformation.......................................148
deformation twins .............................148
electrochemical polishing.........149–150
electrolytic etching ...........................151
electrolytic polishing .........................149
etching techniques and preparation of surfaces..........148–151
etch polishing............................148–149
ghost structures ................................148
grinding..................................148, 149
macroetching.....................................150
microetching.............................150–151
microtoming.......................................148
physical etching ................................151
polishing ..................................148–149
preparation of surfaces.............148–149
smear layers .....................................148
**Zinc chloride**
formula, hazards and properties........196
**Zinc selenide coatings**
for aluminum alloys ............................64
for cobalt alloys..................................80
for vapor deposition ...........................44
**Zinc sulfide coatings**
for aluminum alloys ............................64
for vapor deposition ...........................44
**Zinc telluride coatings**
for aluminum alloys ............................64
for vapor deposition ...........................44
**Zirconium and zirconium alloys**
anodizing..........................................112
chemical-mechanical polishing 109, 110
chemical polishing ....................109, 110
electrochemical polishing .................110
electrolytic etching...........................112
electrolytic polishing ................109, 110
etching techniques and preparation of surfaces..........109–112
etch polishing ...................................110
grinding ...................................109, 110
macroetching....................................111
microetching.....................................111
polishing...................................109, 110
preparation of surfaces.............109–110
ultramilling ......................................109
vibratory polishing...................109, 110

FSC
www.fsc.org
MIX
Papier | Fördert gute Waldnutzung
FSC® C083411

Kolibri 360 GmbH
Ferdinand-Jühlke-Straße 7
99095 Erfurt, Deutschland
produktsicherheit@kolibri360.de